A PHOTOGRAPHIC GUIDE TO THE VEGETATION OF THE SOUTH TEXAS SAND SHEET

Perspectives on South Texas

Sponsored by Texas A&M University–Kingsville

Timothy E. Fulbright, General Editor

A Photographic Guide to the Vegetation of the South Texas Sand Sheet

Dexter Peacock and Forrest S. Smith

TEXAS A&M UNIVERSITY PRESS
College Station

This paper meets the requirements
of ANSI/NISO Z39.48–1992 (Permanence of Paper).
Binding materials have been chosen for durability.
Manufactured in China through Four Colour Print Group.

LIBRARY OF CONGRESS CONTROL NUMBER: 2019944802

ISBN 13: 978-1-62349-782-8 (FLEX: ALK. PAPER)
ISBN 978-1-62349-783-5 (EBOOK)

A list of titles in this series is available at the end of the book.

This book is dedicated

to the private landowners

of the South Texas Sand Sheet.

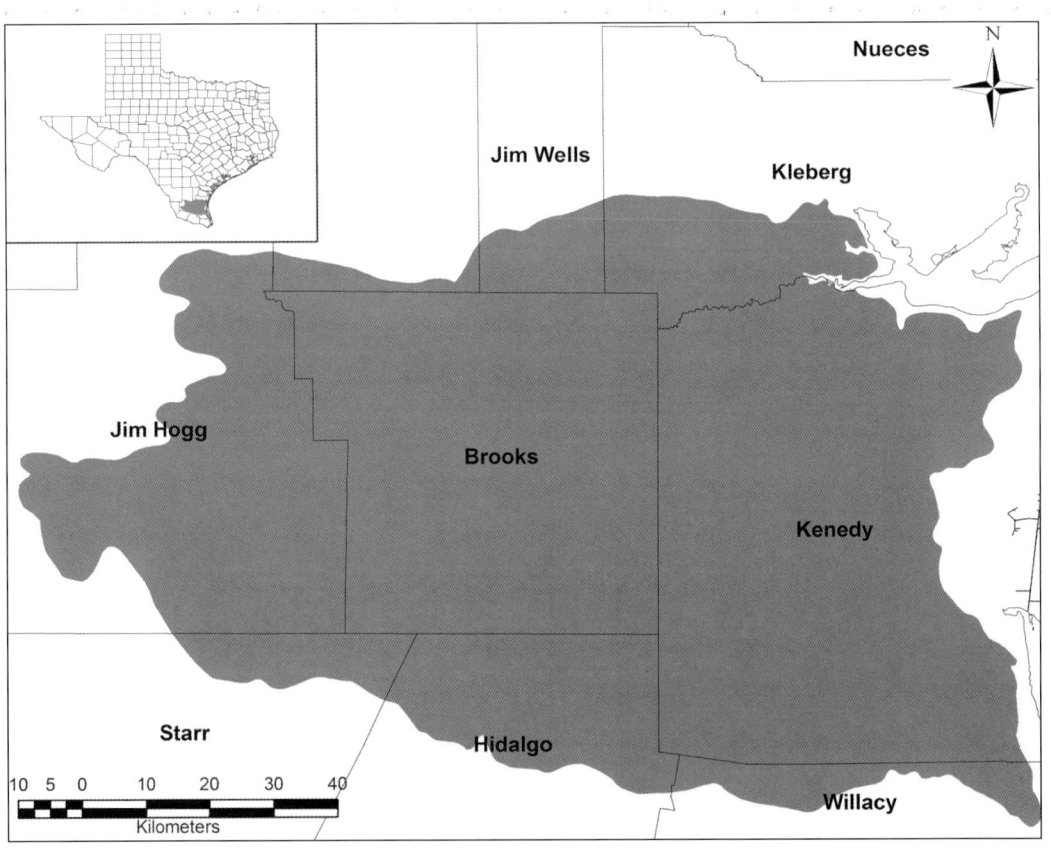

Map of the South Texas Sand Sheet, by the Caesar Kleberg Wildlife Research Institute Geospatial Technologies Lab

Contents

Preface

Forrest S. Smith

My first exposure to the Sand Sheet of South Texas was as a freshman range and wildlife management major at Texas A&M University–Kingsville in the late 1990s. Fresh off the deluge of Hurricane Bret, the Sand Sheet was at its best as I drove down US Highway 77 for the first time en route to a job interview at a hunting camp in the southernmost portion of the region. I made it just past Sarita before I had to stop and look at the plants. Stands of big bluestem, like I had never seen before and have never seen since, abounded in the railroad and highway rights-of-way, and in the adjacent well-managed pastures. Majestic live oak mottes, backed by sand dunes and surrounded by grasslands punctuated with an array of flowers, framed seemingly endless expanses of wild, native land on big ranches with names any country kid, even one reared in the Hill Country, would know—the King, Kenedy, Armstrong, Yturria, and El Sauz, for example. Wildlife—the product of those exceptional habitats—was abundant, especially deer, quail, turkeys, doves, javelinas, and the mysterious nilgai.

For a young student of wildlife and plants, the opportunity to learn here, work here, and play here—*in elemetum* was a perfect description. I had found my Yellowstone, my Everglades, my Amazon. If you learn what you are looking at in terms of its diversity of plants and wildlife, the South Texas Sand Sheet is that level of special place.

What captured my interest then, and probably also lies behind yours now, is that the Sand Sheet landscape and its features—and notably, for our purposes here, its native plants—are unlike any you have seen before or will see anywhere else. The entire region is dominated by native plants and habitat, a very rare occurrence at an ecoregion scale. As a result, for example, you will find northern bobwhites here to hunt in greater numbers than occur anywhere else, past, present, or future. You will see pastures that rival the size of ranches anywhere. In those pastures, you will find endemic plants that grow nowhere else in the world, and the sheer number of different plant species per unit area is greater than it is almost anywhere else in the United States.

One reason you will find those things is that today, you will also find a culture of respect for the land unlike any elsewhere. That perhaps has not always been the case, certainly in times past, when overgrazing and mineral production without cognizance of environmental consequences left many scars on the landscape. Even so, today's land ethic in the Sand Sheet transcends that of the past in the pastures of ranching empires going back eight generations, and in those of newer, hunting- and conservation-focused landowners. What results is a landscape still dominated by relatively undisturbed and diverse native habitats, one not present by accident or by chance. Another unique trait of the South Texas Sand Sheet is the level of focus, effort, and investment in conservation by those ranchers, which you will learn of. That this region still abounds today in native habitats and their resulting native wildlife is in part because of the biological and geographical uniqueness of its location, but also because of its history of ownership and its many dedicated stewards and landowners.

As a student of plants, I have had a steep learning curve, and I was fortunate to be guided by my choice of field of study, where formal classroom training in plants of the region was augmented by friendships with many "sages" of local vegetation. These scientists, ranch managers, and mentors, with their great experience and knowledge, gave me the opportunity to learn quickly. Soon thereafter came my employment at the Caesar Kleberg Wildlife Research Institute (CKWRI), as a student technician of the new South Texas Natives Project. That work, which first entailed collecting native seeds from the Sand Sheet, allowed me to further hone my interest in and knowledge of native plants. Eventually, my passion for their conservation and restoration turned into a career.

*But all along the way in this journey, when it came to learning about the vegetation of the Sand Sheet, I had to battle the fact that there was **no book**.*

While I had the luxury to learn from many over the years, not everyone has the time, avocation, and vocation, or the opportunities and friends that I did. Without that unique background, learning even the common vegetation of the Sand Sheet on my own or with available resources would have been as tough as cactus and barbed wire. While plant books from the Lower Rio Grande Valley, South Texas as a whole, or elsewhere in Texas are available and excellent, they only tangentially cover the breadth of the Sand Sheet's unique flora. In my professional work as a scientist at the CKWRI, when asked to point students and other interested persons toward resources for identifying Sand Sheet vegetation, I was continually frustrated, because no specific text existed. The best one could do was consult a bookshelf's worth of various works from elsewhere, which included some of the plants that might be encountered. Many of my colleagues and I repeatedly mused over doing a plant book specific to the Sand Sheet, but alas, one never got off the ground.

Fast-forward to 2015, around a kitchen table at Dexter Peacock's Sand Sheet ranch in Jim Hogg County. Dexter's frustrations with identifying plants using the available resources had led him to ask me out to look at plants, identify photos he had taken, and work through a great many identification snags. His observations then mirrored mine, and his hobby of identifying the plants on his ranch, and the need for somebody to fill the gap in available resources, became a project.

What followed is a product of no small effort. We spent long days afield, finding and photographing plants; we scoured through hundreds of photographs, his and mine, choosing only photos that "looked" like what the plants really "looked like"; and we wrote draft after draft, crafting descriptions unlike any that anyone had written before in a work like this (unabashedly and on purpose), so that anyone could understand them. That resulted in this book. It is thanks mainly to Dexter that it has arrived. I am beyond glad for his prodding to the finish, and I am very glad I worked with him on a plant book, instead of on the opposite end of a legal matter, which was his profession before ranching.

Our hope all along has been that this book will fill the gap we think we have identified, and that those who open its pages will quickly be able to accurately identify a great deal of the incredible vegetation they will encounter in the South Texas Sand Sheet. Even more, my hope is that it helps the next wide-eyed students who venture south on Highway 77 to learn more quickly, to learn better, and to do even more someday to help wisely manage and conserve the Sand Sheet and areas like it in their future careers. I hope it motivates more landowners like Dexter, as well as the thousands of hunters, ranchers, and guests who come here each year, to own, ranch, hunt, or recreate in such a way that today's Sand Sheet, if not a better one, endures into the next generation and many thereafter.

If in those last endeavors we have some success—and we need to—we will have met our mark. In less than just the two decades since my first exposure to the Sand Sheet, much has changed. Invasive species, notably guineagrass, but also others like natal grass, Wilman lovegrass, and Lehmann lovegrass introduced here from abroad on purpose or by accident, have exploded in distribution and amount. Even tanglehead, a native grass, has changed greatly in behavior, a concern of many. The resulting effects on plant diversity, and ultimately on wildlife and the region's lifeblood, are still playing out, but they concern many of us greatly. Identifying these or future problems, or positive trends in good management, such as large stands of big bluestem, begins with identifying the plants, something I am confident this book will help anyone do.

Other changes to the Sand Sheet, not originating in the plant world but impacting it greatly, have also occurred. Land is being fragmented, and one of the nation's fastest-growing urban areas is creeping north, swallowing parts of the Sand Sheet in its wake. More roads, more utility easements, more houses, and collectively more of the "footprints" of progress are waylaying native plant communities and introducing invasive and nonnative plants. Energy exploration and transport and the resulting scars they can leave are certainly also impacting the Sand Sheet more and more. Not one wind turbine towered above the Sand Sheet I first found not that long ago, and obviously today there are many. Appreciating why any of these concerns *are concerns* starts with knowing what is lost and why it is special, or what it is you may want to restore when the opportunity presents itself. While the list of losses from our human footprint, poor management, or abhorrence of nature often starts with "just" the plants, ultimately the impacts are so much more. In using this book, we hope you will learn what is at stake and learn to appreciate it a little bit more, just as we have had the distinct pleasure of doing in preparing this work.

I look forward to seeing a ragtag, well-used copy of our book on your quail rig or truck dashboard.

Acknowledgments

The authors would like to thank Rowan Companies PLC and the Caesar Kleberg Wildlife Research Institute for providing funding for this book; David Grall, who got this book started; and the many ranchers who gave us access to their ranches, including Jim Gibbs of the San Pablito Ranch; James Myers of the Palangana Ranch, Stephen Burns of the Visnaga Ranch; Todd Johnson of La India Ranch; and Dan English of the San Chicago Lease on the Norias Division of the King Ranch.

Forrest Smith's work on this book was supported in part by the Dan L. Duncan Endowment and by the numerous private landowner-donors to the South Texas Natives Project and through the Caesar Kleberg Partners Program.

And thanks to Beto Trevino of Rancho Lomitas, who supplied many of the Spanish names used in the book.

A PHOTOGRAPHIC GUIDE
TO THE VEGETATION
OF THE
SOUTH TEXAS SAND SHEET

Introduction

Dexter Peacock

The idea for this book began in frustration. I got interested in the amazing vegetation of the South Texas Sand Sheet shortly after I acquired a small ranch in Jim Hogg County. I bought a book on the subject. Then I realized I needed several books, because none of them dealt with more than a couple of the types of plants I was seeing. Before long, I had fourteen books (I will call them the "Botanical Guides"; see the bibliography).

At first I was amazed that there were so many books. That confirmed to me that a lot of people out there were as fascinated as I was with the plants of South Texas. I naively dug into them, believing I could eventually achieve some degree of mastery of this subject. I came to realize, however, that I faced a major obstacle: I wasn't trained in botany. I hadn't taken any botany courses in college. And most of these books assumed that the reader would be at least conversant with botanical terms. Myself, I didn't know a calyx from a rachis.

But, I thought, most of these books have photos of these plants. Surely I can identify what I see in the field by using the Botanical Guides.

Often, I could. But often I couldn't. And it almost always entailed a lot of work.

Part of the problem was their organization; without exception, the Botanical Guides are organized by genus and species, not by appearance of the subject. This makes perfect sense, if you are a botanist and know the scientific organization of species. But if you aren't, you can spend hours thumbing through them page by page, trying to find a photograph that resembles what you see.

Another problem is that the photographs aren't always very good, and they are pretty small in some of the Botanical Guides. Or they were taken in the spring, and you are looking at plants in December. Another realization occurred—photography was not the main concern of the scholars who wrote the Botanical Guides. (In fairness, the later the publication date of the book, the better the photos are).

Over time, I concluded that the Botanical Guides were not intended primarily for people like me—nonbotanists interested mainly in knowing the name of what they find in the field. Many of the Botanical Guides make an effort in this direction, but my experience has been that you have to know something about botany to be sure of your identification in many cases. Not all, but enough to make it pretty frustrating at times.

What, I thought, about doing a book that covered all the common species of vegetation in one volume, so you wouldn't have to tote around several (or all) of these books? What, I thought, about organizing the book differently, putting plants that look alike together, so that the reader could zero in on the identification right away? Say, putting all the plants with yellow flowers in the same section (regardless of genus and species). After all, most of us wouldn't know whether a cowpen daisy was a member of the Asteraceae or the Fabaceae family, but we can all see that it is yellow. Better yet, what about putting all the plants with yellow petals and yellow heads together, next to the section containing plants with yellow petals but brown heads? What about trying to get really good photos of these plants, photos that would themselves permit identification? And accompanying those photos with a brief text in plain English that pointed out visual clues to identification and suggested how to tell plants that resemble each other apart? Since photography has been a serious interest of mine all my adult life, I thought: How hard could that be?

I convinced myself that I was onto something. So, about eight years ago, I began to assemble a collection of plant photographs, now numbering well over a thousand. Slowly, however, I reached the conclusion that it is not possible for a nonbotanist, regardless of his or her degree of diligence, to be sure of his or her identifications of some of these species using the photographs in the Botanical Guides alone.

I had the good judgment to realize I needed professional guidance and to enlist the help of an expert: Forrest Smith, director of the South Texas Natives Project at the Caesar Kleberg Wildlife Research Institute at Texas A&M-Kingsville. Forrest tactfully agreed that I needed help and agreed to become my coauthor. Importantly, he agreed that there is a gap in the literature on this subject that needed to be filled by a book like ours. He has an encyclopedic knowledge of the vegetation of South Texas, and we have now spent many hours together in the field. A number of the photographs in this book are his, including some of the spectacular ones, such as figure 165. Because of his participation, I am confident that the plants pictured in this book are what we say they are.

And now I have learned what a calyx and a rachis are. They are

Weeds and Forbs
Conspicuously Flowered Forbs

YELLOW-FLOWERED FORBS
Partridge Pea
Chamaecrista fasciculata (Michx.) Greene

Partridge pea is one of the iconic plants of the Sand Sheet. It is highly valued by wildlife managers and hunters as a food source for northern bobwhites. Yet it is often mistaken for other similar-looking yellow-flowered plants common to the Sand Sheet. Note (a) its fernlike leaves, and (b) the black stamens in its flowers emanating from red bases. It is a stemmed plant of medium height (up to two feet tall) that usually does not appear until summer in the western Sand Sheet, earlier in the eastern part. It is common on sandy soils throughout, and an important pollinator plant. It is not consumed by livestock.

A close look-alike to partridge pea also appears in the Sand Sheet: woodland sensitive pea, *Chamaecrista calycioides* (DC. ex Collad.) Greene. Since it is difficult if not impossible to distinguish from partridge pea by photograph alone, none is included.

Figure 1. Partridge pea—April

Figure 2. Partridge pea—September

Texas Senna

Chamaecrista flexuosa (L.) Greene var. texana (Buckley) Irwin & Barneby

Texas senna is a low-growing, yellow-flowered, fern-leaved plant that is often mistaken for partridge pea. But its stamens are also yellow, unlike those of partridge pea, which are black, and its leaves are finer. It also grows in thicker bunches, wider than tall. Texas senna usually appears earlier in the spring than partridge pea. It is also a good source of bird food and is often a preferred food plant of northern bobwhites early in the hunting season. It is rarely consumed by livestock.

Figure 3. Texas senna flowers—April

Figure 4. Texas senna—April

Bracted Zornia (*Viperina*)
Zornia bracteata Walter ex J. F. Gmel.

Bracted zornia can be mistaken for Texas senna at first glance, but its flower is bilateral, meaning both of its longitudinal halves are identical. Its leaves are not fernlike and have their own, somewhat unusual four-leaflet arrangement. It is common on sandy loam soils. White-tailed deer often eat the plant.

Figure 5. Bracted zornia—March

Figure 6. Bracted zornia flower and leaves—April

YELLOW PETALS, YELLOW HEADS

Figures 7–26 show thirteen different species that all have both numerous yellow petals and yellow heads (the "head" is the center of the flower, from which the petals emanate). The shapes of the petals are different, however, and so are the leaves.

Cowpen Daisy (*Hierba Amarilla*)
Verbesina encelioides (Cav.) Benth. & Hook. f. ex. A. Gray

Cowpen daisy is a common annual, ubiquitous in the Sand Sheet. Note the fringed, separated, stick-shaped petals. Its leaves are fringed and ovate (egg shaped, with the narrower part at the end) to heart shaped. Livestock avoid grazing the plant, but it is an important seed producer for northern bobwhites and a valuable nectar plant for pollinators, including monarchs during their autumn migration through the area. Cowpen daisy blooms throughout the year.

Figure 7. Cowpen daisy—November

Figure 8. Cowpen daisy—May

Awnless Bushsunflower
Simsia calva (Engelm. & A. Gray) A. Gray

This perennial sunflower also has fringed, ovate leaves, but narrower than cowpen daisy's (see fig. 9 vs. fig. 7). Its corolla is different from those found in our annual species, common and sand sunflower, which are brown. The petals of its ray flowers are dispersed and appear to issue from below the disk flowers (see fig. 9).

It is an excellent forage plant for livestock and a favored food plant of deer, often found within and vining through small shrubs. The seeds are consumed by northern bobwhites. An adapted commercial seed source of bushsunflower, called Venado Germplasm, is available for use in reseeding projects.

Figure 9. Awnless bushsunflower —April

Figure 10. Awnless bushsunflower flower—May

Huisache Daisy

Amblyolepis setigera DC.

Huisache daisy has distinctive petals. They are broad and widely separated and have three or four notches at their ends. Huisache daisy is an early-blooming spring annual that is important for pollinators, and it is frequently eaten by deer. It blooms only in the spring and early summer.

Figure 11. Huisache daisy—February

Texas Groundsel

Senecio ampullaceus Hook.

Texas groundsel's flower heads are distinctively large in relation to the overall size of the corolla, and its flower petals are elliptic (pointed at both ends and wide in the middle—about one-third as wide as long) and often notched at their ends. Its stems are thick and tubular. It is a common, early-blooming wildflower found on very sandy soils. It is usually avoided by livestock but is a good pollinator plant, mainly because it is the harbinger of spring—usually the first blooming wildflower of the growing season. When in bloom, any collection of Texas groundsel will attract many black swallowtail butterflies (note black swallowtail in fig. 13). Flowers are common as early as January.

Figure 12. Texas groundsel —March

Figure 13. Texas groundsel—January

Bristleleaf Dogweed
Thymophylla tenuiloba (DC.) Small

The distinctive characteristics of bristleleaf dogweed are its small stature and spiky, thistle-like leaves (the lanceolate leaves in this photograph do not belong to bristleleaf dogweed—"lanceolate" means narrow, rounded at the base, and sharp and pointed at the end, like the tip of a lance). It is a showy wildflower found on shallow soils. The leaves are eaten by deer. Bristleleaf dogweed blooms throughout the year following rainfall.

Figure 14. Bristleleaf dogweed—April

Squarebud Daisy or Showy Nerveray
Tetragonotheca repanda (Buckley) Small

Squarebud daisy is a unique but common species found on very sandy soils. It is endemic to Texas, meaning it is found nowhere else in the world. It has widely separated yellow petals and corollas with tall, bunchy yellow filaments ("filaments" are the stalks of the stamens, or the rod-shaped protrusions emanating from the head of the flower). Flowered stalks often appear in the same bunch with stalks having dead flowers. A good identifier of squarebud daisy is the four pointed green sepals of its calyx, which are visible even from the top of the flower (see fig. 15). (The "calyx" of a flower is its underside base, and its "sepals" are the multiple bladelike extensions making up the calyx; see fig. 16.) It blooms from late spring through summer.

Figure 15. Squarebud daisy—May

Figure 16. Square-bud daisy flowers and calyx—April

Texas Sleepy Daisy
Xanthisma texanum DC. var. orientale Semple

Texas sleepy daisy is an annual that is endemic to Texas. It grows up to two feet high and has flowers with yellow to orange corollas that open around midday. Plants usually have both dry and open flowers. Its leaves are lanceolate and wrap close to the stems. It is common on sandy loam sites.

Figure 17. Texas sleepy daisy—May

Smallflowered or Texas Dandelion
Pyrrhopappus pauciflorus (D. Don) DC.

Smallflowered dandelion has petals that are sharply notched at their ends, and distinctive black-ended stamens. It is a common winter annual in much of the Sand Sheet. This plant blooms from spring through early summer and occasionally during other periods if there is heavy rainfall.

Figure 18. Smallflowered or Texas dandelion—March

Paperflower

Psilostrophe gnaphalioides DC.

Paperflower is another flower with yellow petals and yellow heads, but its petals are distinctively three lobed. Its leaves are elliptic and hairy, which distinguishes it from most of the others. The blooms in the lower right quadrant of figure 19 illustrate why the flower was named "paperflower"; as they die, the flowers turn white and papery and remain on the stalk for a long time.

Figure 19. Paperflower—June

Broom Groundsel
Senecio riddellii Torr. & A. Gray

Broom Groundsel is a showy but unpalatable native wildflower, found only on very sandy soils. The petals of its corolla are widely separated, which distinguishes it from most of the other yellow-petaled, yellow-headed species, and they grow from an extended bud. It has long, thin, cylindrical leaves that grow upward and form wispy bunches. It is typically avoided by grazers. It generally blooms throughout the year with good moisture.

Figure 20. Broom groundsel—May

Orange Zexmenia
Wedelia acapulcensis Kunth var. hispida (Kunth) Strother

Orange zexmenia is a common, perennial subshrub found on tighter soils. Its flowers are more orange than yellow, which is a quick way to narrow down the search for it among plants in this subsection. It is a bushy plant with dense stem growth; dead stems from the previous year protrude and can make the plants appear overgrown and scraggly. It has toothed, lanceolate leaves. It provides good forage for livestock and wildlife, especially white-tailed deer, and produces seeds eaten by quail. It is hardy and blooms following a rainfall. A commercial seed selection called Goliad Germplasm is an excellent addition to rangeland seeding mixes for the Sand Sheet.

Figure 21. Orange zexmenia flowers—May

Figure 22. Orange zexmenia—May

Roughpod Bladderpod

Lesquerella lasiocarpa (Hook. ex A. Gray) S. Watson var. berlandieri (A. Gray) Payson

There are a number of bladderpods in the Sand Sheet, but roughpod bladderpod is by far the most common. Bladderpods are hard to tell apart by their corollas alone, because all have four yellow, rounded petals. Their leaves are probably better keys to identification. Roughpod bladderpod's leaves are basically lanceolate but are deeply and sharply lobed. Its fruits are flattened, which is one additional way to tell this bladderpod from the others.

Its seeds are eaten by northern bobwhites, and the whole plant is eaten by deer. It is one of the first wildflowers to bloom each spring and can create expansive carpets of yellow blossoms.

Figure 23. Roughpod bladderpod —February

Silver Bladderpod
Lesquerella argyraea (A. Gray) S. Watson

Silver bladderpod flowers are small, yellow, and four petaled. Its leaves are lanceolate, thick, and pubescent (fuzzy), and its fruits are round but elongated, unlike the fruits of roughpod bladderpod. Deer eat the leaves.

Bladderpods are important winter annuals, often the first flowers to bloom each spring. They are frequently eaten by cattle and produce copious seeds eaten by doves and quail.

Figure 24. Silver bladderpod—March

Scrambled Eggs
Corydalis micrantha (Engelm. ex A. Gray) A. Gray

Scrambled eggs is one of the earliest-blooming spring wildflowers in the Sand Sheet. It is a winter annual, endemic to Texas, growing up to a foot in height. The clustered yellow flowers curl up and around themselves, making the bunch resemble scrambled eggs. A peculiarity that helps in identification is that the stalks of each flower petal attach to the stems at the midpoint of the flower, giving the flower a structure like an arm holding a megaphone. The leaves of the plant look like those of parsley. Both the leaves and seeds are reported to be eaten by wildlife.

Figure 25. Scrambled eggs —March

Figure 26. Scrambled eggs—March

YELLOW PETALS, BROWN HEADS
Common Sunflower (*Mirasol*)
Helianthus annuus L.

Figure 27. Common sunflower bunch—June

Figure 28. Common sunflower flowers—August

Sand Sunflower

Helianthus praecox Engelm. & A. Gray ssp. runyonii (Heiser) Heiser

Common sunflower and sand sunflower flowers look a great deal alike (compare figs. 27 and 29), but their leaves are distinctive. Common sunflower leaves are lanceolate, ovate (meaning wide and rounded at the base and narrowing in a regular curve to form a point or a slightly rounded end), or broadly triangular, whereas sand sunflower leaves are uniformly triangular, rippled, lobed, and pointed and grow in clusters around the stems, often all the way up to the corollas. Another difference is that common sunflower is much taller (though not nearly as tall as silverleaf sunflower; see fig. 31) and single stalked, with leaves emerging from the main stem except at the apex, whereas sand sunflower is characteristically shorter and branching throughout.

Common sunflower is often found on old farmland and disturbed sites and is one of the most preferred food sources of mourning doves and northern bobwhites during hunting season. Cattle will consume the plant when available. It can reach above-head height by summer in years of abundant rain.

Sand sunflower is a fair seed producer for quail and is generally avoided by livestock. It is usually found on very sandy soils, and in disturbed sites in generally intact native plant communities. It is an important insect and pollinator plant over much of the Sand Sheet. It is endemic to Texas.

Sunflowers are so named, of course, because their corollas turn during the day to face the sun as it moves in its arc.

Figure 29. Sand sunflower bunch—May

Figure 30. Sand sunflower flower—September

Silverleaf Sunflower
Helianthus argophyllus Torr. & A. Gray

While silverleaf sunflower is obviously a member of the sunflower family, it is easily distinguishable from common and sand sunflowers because of its size. Silverleaf sunflower is the tallest sunflower, and the tallest herbaceous plant in the Sand Sheet, growing to heights exceeding ten feet (note the almost-hidden fence pole in fig. 31, which is about four feet tall). The leaves are ovate and heart shaped and can be bigger than your hand. Both the leaves and stems are covered in small hairs, which gives the plant a fuzzy blue-green color. It usually makes its appearance at the end of summer, later than common or sand sunflower. Its seeds are eaten by a variety of birds, though they are not a favored food of northern bobwhites or doves.

This sunflower grows on very sandy soils, especially in disturbed areas. It is considered invasive and problematic by many ranchers. It can dominate heavily grazed pastures and outcompete almost all other vegetation seasonally, especially after copious summer rains.

◄ Figure 31. Silverleaf sunflower stalks—September

▲ Figure 32. Silverleaf sunflower corollas and leaves—September

Tickseed
Coreopsis tinctoria Nutt.

Tickseed is a stalky plant with dense linear leaves. Because of its over-all bushy shape, it could be confused with orange zexmenia (figs. 21 and 22), but the latter has orange heads whereas tickseed has brown heads. It is very common on tight soils. The leaves are eaten by deer and cattle.

Figure 33.
Tickseed—May

Stiff Greenthread
Thelesperma filifolium (Hook.) A. Gray

Stiff greenthread has yellow to gold ray flowers with reddish-brown centers. As with several of the other yellow-flowered, brown-headed flowers in this book, some of its anthers (the pollen-bearing heads of the filaments; filaments and anthers make up the stamens) are yellow, which can be misleading. Stiff greenthread, however, is much shorter in stature than common or bush sunflower, and its rays (or "petals") are blunt and lobed, whereas the petals of both the sunflowers are somewhat pointed. Its flowers grow at the ends of long, branched, leafless stalks.

Figure 34. Stiff greenthread—May

American Snoutbean
Rhynchosia americana (Houst. ex Mill.) M. C. Metz

American snoutbean is a low-growing plant with distinctive, broadly rounded, deeply veined leaves. It produces yellow flowers from spring through fall. It is good forage for livestock and white-tailed deer, and a valuable seed producer for northern bobwhites. In figure 35, it has vined up on grass stalks, which is not its normal configuration. It is most often encountered with its leaves close to the ground, as in figure 36.

Figure 35. American snoutbean flower—September

Figure 36. American snoutbean leaves—March

ROUNDED COROLLA, YELLOW FLOWERS

Flax, Laredo flax, showy evening primrose, and shrubby woodsorrel all share a similar corolla configuration. The petals are wide and abutting or overlapping, giving the corollas an overall cupped and rounded appearance. The differences in their coloration and stamens, however, make them relatively easy to tell apart.

Flax
Linum alatum (Small) Winkl.

Flax corollas have reddish centers, made up of many radiating individual lines of color, and yellow stamens. Flax is a common spring flower across much of the Sand Sheet.

Figure 37. Flax—March

Laredo Flax
Linum elongatum (Small) Winkl.

Laredo flax corollas also have easily recognizable coloration, with concentric circles of red, white, and red issuing from the center. Laredo flax is a common spring flower in the Sand Sheet.

Figure 38. Laredo flax—March

Showy Evening Primrose
Oenothera grandis (Britton) Smyth

Evening primrose is also a common annual wildflower, with bright yellow petals and very recognizable yellow stamens that branch out sideways and crisscross each other, making it easy to identify.

Figure 39. Showy evening primrose—March

Shrubby Woodsorrel
Oxalis frutescens L. ssp. augustifolia (Kunth) Lourteig

Shrubby woodsorrel has a five-petaled yellow flower. Its leaves are broadly elliptical. It is usually found as small shrubby individual plants growing together in dense grass stands. Usage of the plant by wildlife and livestock is undocumented, but it is usually absent from overgrazed sites.

Figure 40. Shrubby woodsorrel—September

Camphorweed

Heterotheca subaxillaris (Lam.) Britton & Rusby

Camphorweed is a fairly tall weed (three feet or higher at maturity) that has leaves at regular intervals all the way up the stems. It also has multiple flower buds on each stem, which distinguishes it from Florida waltheria (fig. 42). It gets its name from the camphor smell it puts out, which makes it highly unpopular with quail hunters because its strong odor tends to overpower hunting dogs' sense of smell.

It is a common biannual forb that forms monocultures on very sandy soils. It provides excellent cover for northern bobwhites and produces seeds they sometimes consume. However, it limits grass growth and can occupy a disturbed site for many years following overgrazing.

Figure 41. Camphorweed— November

Florida Waltheria (*Hierba de Soldado, Uhaloa*)
Waltheria indica L.

Florida waltheria resembles camphorweed in the winter after the flowers have dropped off, leaving both species with brown balls at the ends of their stems. As can be seen in figure 41, however, camphorweed's flower stems have a branched structure and multiple flower buds, whereas Florida waltheria has just one bud per stem. Florida waltheria is one of the most common native forbs on sandy loam soils in the Sand Sheet. Cattle generally avoid grazing this plant.

Figure 42. Florida waltheria—April

South Texas False Cudweed
Pseudognaphalium austrotexanum G. L. Nesom

South Texas false cudweed is extremely prolific following wet summers in the Sand Sheet. It has white and yellow flowers in the fall, and thick growths of slightly lobed, linear leaves. In winter, its flowers fade to brown and cream, then shrivel to brown buds. Most notably, the thick array of leaves surrounding the stems die, turn gray and brown, and droop heavily down the seed stems (see fig. 44). It is typically avoided by cattle. It becomes highly visible in winter after a wet summer because its stalks, with brown buds, remain after other plants have disappeared.

Figure 43. South Texas false cudweed—early November

Figure 44. South Texas false cudweed—late December

Common Broomweed
Amphiachyris dracunculoides (DC.) Nutt.

Common broomweed, often referred to as snakeweed, looks like a small, sparse bush with a many-branched, delicate-looking stem structure and small yellow flowers. It is prolific in wet years. It has poor grazing value but provides good cover and seed production for northern bobwhites.

Figure 45. Common broomweed—November

Figure 46. Common broomweed —November

Short Gland or South Texas Clammyweed
Polanisia erosa (Nutt.) Iltis ssp. breviglandulosa Iltis

Short gland clammyweed is an annual that is endemic to Texas. Its yellow flowers have spectacularly long yellow stamens that protrude from the heads. Its leaves are linear and its stems are pubescent. It is usually found in high-quality, lightly grazed grasslands on very sandy soils. Clammyweeds produce seeds relished by doves and northern bobwhites.

Figure 47. Short gland or South Texas clammyweed—April

Indian Rushpea
Hoffmannseggia glauca (Ortega) Eifert

Indian rushpea's most important visual key is its leaves. They are basically round and grow opposite each other up and down the entire stalk, giving a strikingly symmetrical appearance. It has small yellow flowers. Another plant in this book, roundleaf tephrosia (fig. 103), has a similar leaf structure, but Indian rushpea's leaves are small, one-half to one inch in diameter, whereas roundleaf tephrosia's leaves can be two inches in diameter and are not consistently opposite. Roundleaf tephrosia also has purple flowers.

Indian rushpea is found infrequently on sandy loam soils of the Sand Sheet. Deer eat the leaves.

Figure 48. Indian rushpea—March

Figure 49. Indian rushpea flowers—March

Mexican Bonebract
Sclerocarpus uniserialis (Benth.) Hemsl. var. austrotexanus B. L. Turner

Mexican bonebract is a showy perennial, with widely separated yellow flowers. It has yellow to orange anthers on green filaments that look like small stars with curly points. Its leaves are mainly triangular and deeply toothed. It is common in both open and shady areas of the Sand Sheet. It is an excellent pollinator plant. It is usually found in protected areas or on ranches with no livestock. Its value for livestock and wildlife is undocumented.

Figure 50. Mexican bonebract— September

Seaside Goldenrod
Solidago sempervirens L.

Seaside goldenrod is a tall wildflower with heavy, elongated fronds of yellow flowers (or "inflorescences") that often cause it to bend over in the fall when it is in full bloom. It is an important nectar plant for pollinators, including monarchs. It is usually avoided by cattle.

Figure 51. Seaside goldenrod—October

WHITE-FLOWERED FORBS

Black Foot Daisy or Hoary Blackfoot
Melampodium cinereum DC. var. ramosissimum (DC.) A. Gray

Figure 52. Black foot daisy—March

Lazy Daisy or Dozedaisy
Aphanostephus skirrhobasis (DC.) Trel. var. kidderi (S. F. Blake) B. L. Turner

The most obvious difference between these two daisies is their petal arrangement. Black foot daisy's petals are notched and widely separated, appearing in some cases to be paired. Black foot daisy's leaves are linear and lightly notched. Lazy daisy petals are pointed and regularly dispersed around the corolla, and its leaves are linear to spatulate (meaning narrow at their bases and wide and round at their tips), and pubescent. Lazy daisy is found on sandy soils in open areas. It is frequently eaten by deer and livestock. Black foot daisy is usually found on tight soils, overlying caliche.

Figure 53. Lazy daisy—May

Spiny Pricklypoppy
Argemone sanguinea Greene

Spiny pricklypoppy has cupped white flowers with yellow centers. Its leaves are triangular but have deep lobes and distinctive white veins. The plant is prickly and stings to the touch. It is a common winter annual found in disturbed soils. It produces copious seeds that are eaten by doves and northern bobwhites.

Figure 54. Spiny pricklypoppy—March

Texas Bullnettle (*Mala Mujer*)
Cnidoscolus texanus (Müll. Arg.) Small

Bullnettle's flowers are white and have five widely separated petals with white centers and yellow bases. It is recognizable by the white stinging hairs on its leaf stems. Its leaves are palmate (meaning separating like a palm and fingers from a single point) and lobed.

Figure 55. Texas bullnettle—March

False Ragweed (*Cicutilla*)
Parthenium confertum A. Gray

False ragweed has tiny white flowers and broad double-lobed leaves. It has a strong smell, which sometimes causes confusion with western ragweed. It is common on tighter soils. It has poor grazing and wildlife value but is a good plant for pollinators, and a valuable colonizer after disturbance.

Figure 56. False ragweed—April

Mistflower (*Crucita*)
Chromolaena odorata (L.) R. M. King & H. Rob.

The chief visual characteristic of mistflower is its dense collection of slender stems bowing upward to small, white or purple flowers. It can be mistaken for three-lobed florestina (see fig. 58), but three-lobed florestina is not bushy and its flower stems are less prominent. The leaves of mistflower are lanceolate and lobed. It is often found growing under brush canopies. Mistflower is an important butterfly plant in the Sand Sheet, and white-tailed deer often browse the leaves.

Figure 57. Mistflower—October

Three-Lobed Florestina
Florestina tripteris DC.

Three-lobed florestina is a spindly-looking plant about two feet high that has small white flowers and three ovate/elliptic leaves growing from the same node ("three lobed"). It is common on sandy loam sites in some areas of the Sand Sheet. It has poor grazing value but is an excellent pollinator plant.

Figure 58. Three-lobed florestina—June

Figure 59. Three-lobed florestina—May

Scorpion's Tail (*Cola de Alacrán*)
Heliotropium angiospermum Murray

Scorpion's tail is a beautiful heliotrope with parallel, coiling rows of flowers that become progressively smaller as they extend outward from the stem. The corollas are white with yellow centers, and the leaves are ovate. The plant prefers shady locations. Heliotropes are good plants for pollinators and are eaten by deer.

Figure 60. Scorpion's tail—April

Coastal Plains Heliotrope
Heliotropium racemosum Rose & Standl.

Figure 61. Coastal plains heliotrope—September

Texas Heliotrope
Heliotropium texanum I. M. Johnst.

Both of these heliotropes are low-growing plants with very small flowers (note the pop-top at the bottom of fig. 61). Their corollas look very much alike—five tiny white petals with yellow centers—and both have pubescent (fuzzy) leaves. Their leaves are slightly differently shaped, in that Texas heliotrope leaves are elliptic while coastal plains heliotrope leaves are more lanceolate. Generally, however, the best visual key is that coastal plains heliotrope grows in small, low-growing clumps while Texas heliotrope tends to produce upward-growing stems with bunches of flowers at their tops. Coastal plains heliotrope is endemic to Texas.

Heliotropes are eaten by white-tailed deer and are important pollinator plants.

Figure 62. Texas heliotrope—June

Drummond's Snakecotton
Froelichia drummondii Moq.

Drummond's snakecotton is a stemmy plant that has elongated clusters of small white flowers that grow upward and close to the main stem. They have a fuzzy, or cotton-like, appearance. Drummond's snakecotton has no particular value to wildlife other than providing cover, and habitat to insects. It is common on very sandy soils.

Figure 63. Drummond's snake-cotton—May

Figure 64. Drummond's snakecotton—December

Rabbit Tobacco or Evax

Evax verna Raf. var. verna

Rabbit tobacco is a low-growing, hairy plant with stems that spread more horizontally than vertically. Its leaves are spatulate, and its flowers occur in closely grouped white clusters.

Figure 65. Rabbit tobacco or evax—April

Zizotes Milkweed
Asclepias oenotheroides Cham. & Schltdl.

Zizotes milkweed is a truly striking plant that has groupings of small white hoods that surround the yellow centers of each flower and extend upward like the five fingers of a hand touching each other at the fingertips. It is infrequent in the Sand Sheet. It is an important host plant for monarchs during their spring and fall migration through the area, and an excellent pollinator plant overall. At maturity, pods harbor flat, disklike seeds attached to filaments called floss; when they burst, they release the seeds to be distributed by the wind.

Figure 66. Zizotes milkweed—April

Emory's Milkweed
Asclepias emoryi (Greene) Vail

Emory's milkweed also has an interesting flower arrangement, with green lobes drooping downward from yellow and white centers. Its leaves are ovate but narrow, with prominent toothing and an acute concave cupping. Emory's milkweed is infrequent on red sandy loam soils in the Sand Sheet. It is a host plant for monarchs, and an excellent pollinator plant. As in all milkweeds, the seeds are borne in pods and are attached to floss for distribution across the landscape by wind.

Figure 67. Emory's milkweed—March

Woolly White or Old Plainsman
Hymenopappus artemisiifolius DC. var. riograndensis B. L. Turner

Figure 68. Woolly white—April

Figure 69. Woolly white flower head—April

Wild Buckwheat
Eriogonum multiflorum Benth.

Woolly white and wild buckwheat are somewhat similar in appearance, but the clusters of wild buckwheat flowers are flattish (see fig. 71), and its triangular leaves grow close to the stem and extend high up the stems. Woolly white flower clusters have an overall hemispherical shape (see fig. 68). Woolly white's leaves are deeply lobed and thistle-like. Woolly white usually appears early in the spring, while wild buckwheat usually appears in midsummer. In winter, wild buckwheat seedheads retain their shape but turn a dark reddish-brown (see fig. 72). Both plants are common on very sandy soils and provide good cover for wildlife. Both are good pollinator plants.

Figure 70. Wild buckwheat—September

Figure 71. Wild buckwheat—October

Figure 72. Wild buckwheat—November

BLUE-FLOWERED FORBS

The species of blue-flowered plants that are common in the Sand Sheet are all easily distinguishable from each other by the appearance of their flowers, as evidenced by figures 73–78. They are, without exception, beautiful examples of the flora on the Sand Sheet.

Drummond's Skullcap
Scutellaria drummondii Benth. var. drummondii

Drummond's skullcap has a unique (at least in the Sand Sheet) flower structure. The corolla is bilateral (meaning longitudinally symmetrical—both longitudinal halves are identical) and brilliant blue. There are five petals and a tongue of white stamens drooping downward, with blue dots emanating from the center. The leaves are pubescent, ovate, and cupped. It is a showy wildflower found on very sandy soils, and it is eaten by deer and livestock.

Figure 73. Drummond's skullcap— March

Sandy Land Bluebonnet
Lupinus subcarnosus Hook.

Sandy land bluebonnet is endemic to Texas and is common on very sandy soils. Its petals are almost all blue, in contrast to those of Texas bluebonnets, which are blue and white. Texas bluebonnets are found over most of the rest of the state. All bluebonnets in Texas are considered to be the state flower. Its seeds are eaten by northern bobwhites.

Figure 74. Sandy land bluebonnet—March

Whitemouth Dayflower
Commelina erecta L.

Whitemouth dayflower is another unforgettable blue flower of the Sand Sheet. It has two broad, round blue petals above that curve backward, and two smaller, cupped, white or light blue petals below the darker blue petals, with small yellow stamens in the center of the origins of both. The corolla is bilateral. Upon close observation, it can be seen that there are also two sets of blue stamens, one in each lower cup, that make graceful reverse curves inward toward the center of the lower cups. Its leaves are long and linear and are curved inward.

It is an important wildlife and livestock plant throughout the Sand Sheet. Its seeds are eaten by northern bobwhites.

Figure 75. Whitemouth dayflower—March

Blue Curls
Phacelia congesta Hook.

Blue curls has tiny blue flowers and is usually found in the shade of grasses or other plants. Note the tall (in relation to the size of the corollas) filaments with yellow anthers. Its leaves are broadly ovate and double lobed. (The large, obovate [meaning egg shaped, broader at the end], toothed leaves in fig. 76 belong to another plant, in whose shade blue curls is flourishing.) Blue curls is one of the first wildflowers to bloom each spring in the Sand Sheet. It is often found in or near the canopy of brush or trees.

Figure 76. Blue curls—April

Sand Phacelia
Phacelia patuliflora (Engelm. & A. Gray) A. Gray var. austrotexana J. A. Moyer

Sand phacelia is a common annual with trailing stems—that is, it grows in a sprawling manner with stems hugging the ground. It resembles blue curls and is a member of the same family, but its flowers are larger and have prominent white centers, unlike those of blue curls. It is common throughout the Sand Sheet and is a prolific early bloomer each spring.

Figure 77. Sand phacelia—March

Blue-Eyed Grass
Sisyrinchium langloisii Greene

Blue-eyed grass has six rounded petals, and corollas with yellow centers. Its leaves are tall, thinly lanceolate, and spiked. It is a striking wildflower that blooms early in the spring and is an important pollinator plant—note the bee in figure 78.

Figure 78. Blue-eyed grass—March

PINK- AND PURPLE-FLOWERED FORBS

The pink- and purple-flowered plants of the Sand Sheet are easily recognizable from the photos in figures 79–107 and really need no extensive description.

Purple Peat Leaf or Purple Nymph
Alophia drummondii (Graham) R. C. Foster

This flower has many names, and it is one of the most striking wildflowers in the Sand Sheet. It blooms throughout the year following rain.

Figure 79. Purple peat leaf—April

Carolina or Blue Larkspur
Delphinium carolinianum Walter

Carolina larkspur has blue flowers that hang from the stem in tubes and then blossom outward, resembling the shape of saxophones. Its flower heads can vary from lavender to white to blue. This flower is hard to misidentify. It prefers sandy soils.

Figure 80. Carolina larkspur—April

Silverleaf Nightshade (*Trompillo*)
Solanum elaeagnifolium Cav.

Silverleaf nightshade has five-petaled, star-shaped, blue flowers with conspicuous yellow stamens in the center. Its leaves are narrowly oblong and slightly lobed. It is common on disturbed and degraded sites but is a good seed producer for northern bobwhites.

Figure 81. Silverleaf nightshade—September

Sandbell

Nama hispidum A. Gray

Sandbell has tiny purple flowers (note the coin at the bottom right of fig. 82) and is a low-growing plant that spreads horizontally on reddish, hairy stems. Its leaves are elliptic. It is common on very sandy soils.

Figure 82. Sandbell—May

Meadow Pink
Sabatia campestris Nutt.

Meadow pink is a low-growing plant with five-petaled purple flowers that have star-shaped yellow markings at the center of the corollas. One of its popular names is "Texas star."

Figure 83. Meadow pink—April

Winecup

Callirhoe involucrata (Torr. & A. Gray) A. Gray var. lineariloba

Winecup is common in spring over much of Texas. It is a member of the mallow family, other examples of which are shown in figures 116 and 117. Winecup can carpet the western Sand Sheet following wet winters. It grows from an underground bulb and forms vines on the soil surface.

Figure 84. Winecup—March

Wild Petunia

Ruellia nudiflora (Engelm. & A. Gray) Urb. var. runyonii (Tharp & F. A. Barkley) B. L. Turner

Wild petunia has purple flowers with five petals that form a tube as they reach the stem. Its leaves are ovate, toothed, and somewhat rippled. Its leaves are eaten by deer, and the seeds are eaten by northern bobwhites.

Figure 85. Wild petunia—June

Texas Toadflax

Nuttallanthus texanus (Scheele) D. A. Sutton

Texas toadflax grows on stems about two feet tall at maturity. It has five-lobed, bilateral corollas. When flowering, its buds droop downward, positioning its corollas more or less vertically and its largest three petals like the base and arms of a cross. Its stalks grow from a cluster of leaves (not pictured) at its base. It is a common early bloomer, often flowering as early as January.

Figure 86. Texas toad-flax—March

Pink Evening Primrose (*Amapola del Campo*)
Oenothera speciosa Nutt.

Pink evening primrose has four broad, light pink petals. Like the other member of the evening primrose family illustrated in this book (showy evening primrose, fig. 39), pink evening primrose has highly distinctive four-filament yellow stigmas that are horizontal to the head, forming an X pattern when viewed from above. Showy evening primrose, however, has yellow flowers.

Figure 87. Pink evening primrose—April

Phlox

Phlox drummondii Hook.

Phlox is a diminutive five-petaled flower with ten white petal-shaped markings at the center. Its leaves are small and narrowly ovate (the long, spear-shaped leaves in fig. 88 belong to a nearby whitemouth dayflower). Phlox is also a common spring bloomer, found on very sandy soils.

Figure 88. Phlox—April

Gray's Milkpea
Galactia heterophylla A. Gray

There are several milkpeas in South Texas. Hoary milkpea is grouped in this book with inconspicuously flowered plants (fig. 130). Gray's milkpea has a bilateral, violet to purple flower and oblong leaves. Scarlet pea, another plant similar in appearance, has a red to pink flower (fig. 113). Gray's milkpea is a perennial found on sandy loam sites. Its leaves are eaten by deer, and the seeds are consumed by northern bobwhites.

Figure 89. Gray's milkpea—April

Gaura

Oenothera xerogaura W. L. Wagner & Hoch

Gaura is easily recognizable, with a tall, single stalk and pink, round-petaled flowers extending outward from prominent stems in a single direction. It is a showy spring wildflower and is important for pollinators and bees. Several species occur in the Sand Sheet.

Figure 90. Gaura—March

Texas Thistle
Cirsium texanum Buckley

Texas thistle is common and prolific in some years in the Sand Sheet. The plants are tall (about three feet) and often have dead blooms mixed with the flowering blooms (see fig. 92). The clustered pink blooms are a good nectar source for pollinators.

Figure 91. Texas thistle flower—May

Figure 92. Texas thistle—April

Sensitive-Briar
Mimosa latidens (Small) B. L. Turner

Texas thistle and sensitive-briar flowers resemble each other from a distance. But sensitive-briar flower heads have yellow stamens (see fig. 93), and Texas thistle flower heads are completely purple (see fig. 91). Also, Texas thistle is tall and upright, and sensitive-briar is a ground-hugging vine. Sensitive-briar is so called because its leaves curl and close if touched.

Figure 93. Sensitive-briar flower—April

Figure 94. Sensitive-briar— April

Shrubby Beebalm
Monarda fruticulosa Epling

Shrubby beebalm is a plant of medium height (about two feet high), with corollas of alternating layers of white and purple petals. It grows in bunches and has slightly toothed linear leaves. It is sometimes confused with western ragweed (fig. 151) in the winter when the flowers are no longer present. Western ragweed, however, has many-lobed leaves, whereas those of shrubby beebalm are narrowly linear and look "stringy."

Figure 95. Shrubby beebalm—April

Figure 96. Shrubby beebalm flowers—May

Spotted Beebalm
Monarda punctata L.

Spotted beebalm flowers very closely resemble shrubby beebalm flowers (fig. 98 vs. fig. 96), and especially those of horsemint (or lemonmint, as it is also called), another plant found in the Sand Sheet. All three are members of the same family. Shrubby beebalm is a bushy plant, however (see fig. 95), whereas horsemint and spotted beebalm usually grow in stalks. (Spotted beebalm flowers and horsemint flowers are too nearly identical to warrant a separate photo, with the possible exception of the color of the flower petals. Spotted beebalm flowers tend toward white whereas horsemint flowers tend toward pink, but this too can vary with the time of year). All have a pungent, somewhat minty or lemony odor; they are avoided by livestock but are good plants for pollinators and have showy flowers.

Figure 97. Spotted beebalm stalks—May

Figure 98. Spotted beebalm flower—May

Sand Palafoxia

Palafoxia hookeriana Torr. & A. Gray

Figure 99. Sand palafoxia—September

Figure 100. Sand palafoxia flowers—September

Texas Palafoxia
Palafoxia texana DC.

Both sand palafoxia and Texas palafoxia are tall (although sand palafoxia is much taller—up to about seven or eight feet, whereas Texas palafoxia is about three to five feet tall) and have pink flowers. There is an easy way to tell palafoxias apart, however, even when they are not fully grown: the corollas of sand palafoxia have heads and petals; the corollas of Texas palafoxia are roundish, uniformly flowered, and clustered (compare figs. 100 and 102). Also, sand palafoxia has prominent curled, lanceolate leaves all the way up the stem, and the stems are sticky to the touch. Both flower in the fall, not the spring. Sand palafoxia is endemic to Texas.

Figure 101. Texas palafoxia—November

Figure 102. Texas palafoxia flowers—December

Roundleaf Tephrosia
Tephrosia lindheimeri A. Gray

Roundleaf tephrosia is an endemic species that is easily recognizable because of its distinctive leaf arrangement; it has round but elongated leaves, mostly opposite, that give the plant a uniform, dotted appearance from a distance. Its leaf structure is similar to that of Indian rushhpea (fig. 48), but its leaves are much larger than those of Indian rushpea and are somewhat elongated. Its flowers are purple and bilateral. Roundleaf tephrosia leaves are eaten by deer and cattle, and the seeds are eaten by northern bobwhites.

Figure 103. Roundleaf tephrosia—April

Chisme or Shaggy Portulaca
Portulaca pilosa L.

Chisme is perhaps most easily recognizable by its leaves, which are hairy, linear, and succulent (meaning rounded or oval in cross section, instead of flat). It has five petals and yellow stamens. The plant is an important food for many Sand Sheet animal and bird species (note the apparent bite taken out of one of the leaves in the photograph). It is found mainly on the western portion of the Sand Sheet.

Figure 104. Chisme or shaggy portulaca—August

Rio Grande Clammyweed
Polanisia dodecandra (L.) DC. ssp. riograndensis Iltis

Rio Grande clammyweed is an annual of medium height (about two feet) with unusual bushy-looking pink flowers. One visual key is the long stamens that extend well past the corollas. A second visual key is the pods or seed capsules that radiate upward around the corollas. Its leaves are widely elliptic to ovate. It is not grazed by cattle or deer but produces large amounts of seeds consumed by doves and northern bobwhites. A commercial seed source of the plant, called Zapata Germplasm, exists for use in restoration efforts, game bird food plots, and pollinator plantings.

Figure 105. Rio Grande clammyweed—June

Jimson Weed
Datura wrightii Regel

Jimson Weed flowers are almost unmistakable—five interlocking purplish-pink petals with curled, pointed tips. The problem here is that the flowers only open at night or on foggy days. Its leaves are large, ovate, lobed, and deeply veined. It is a poisonous plant.

Figure 106. Jimson weed—September

Sand Verbena or Heart's Delight
Abronia ameliae Lundell

Sand verbena is an uncommon but well-known perennial of the central portion of the Sand Sheet. The pink golf ball–sized flower clusters are unmistakable. Sand verbena is endemic to Texas. The plant is usually found in loose sandy soils, often near live oak mottes or on rolling dune land, on relatively undisturbed native rangelands. The plant is often associated with the Falfurrias area, where it is most common, and where it is affectionately called "Heart's Delight" by locals.

Figure 107. Sand verbena—March

RED- AND ORANGE-FLOWERED FORBS

Three plants with red and yellow petals are common in the Sand Sheet. As the following photographs illustrate, however, they have different looks.

Mexican Hat or Upright Prairie Coneflower
Ratibida columnifera (Nutt.) Wooton & Standl.

Mexican hat is easily recognizable from its cone-shaped flower head with red petals at the bottom of the cone that have yellow terminal margins. It is a weedy, common forb found on tighter soils throughout the Sand Sheet. It is a good plant for pollinators, but it does not provide much in the way of forage for livestock or wildlife. A very similar species, *R. peduncularis* (naked Mexican hat), is common on very sandy soils of the Sand Sheet (see fig. 109).

Figure 108. Mexican hat—April

Figure 109. Naked Mexican hat—May

Nueces Greenthread
Thelesperma nuecense B. L. Turner

Nueces greenthread is a striking plant with red petals with yellow terminal margins. In this respect, it is similar to Mexican hat, but the shape of the corollas is different (see fig. 109). Nueces greenthread has a red hemispherical flower head with tiny yellow stamens in the center. Stiff greenthread, in figure 34, has a solid yellow corolla.

It is a common native wildflower on very sandy soils. This species is poor forage for livestock but an important source of diversity, as it benefits the arthropods and wildlife that consume it. Nueces greenthread is endemic to Texas.

Figure 110. Nueces greenthread—May

Indian Blanket
Gaillardia pulchella Foug.

Indian blanket has red and yellow petals with three distinct lobes, and the petals sometimes have gaps between them. The flower head is also red with yellow anthers. It is one of the most common and prolific wildflowers of the Sand Sheet. It is fair forage for livestock and wildlife, and an important source of nectar for a wide range of pollinators, especially butterflies.

Figure 111. Indian blanket flower—April

Figure 112. Indian blanket—April

Texas Paintbrush
Castilleja indivisa

Texas paintbrush is an iconic Texas plant that can cover large fields in the spring. A showy, annual wildflower, it is found on very sandy soils, especially in areas protected from livestock, and often seen in the mowed areas of highway right-of-ways where competition with grass is reduced. The plant is often mistakenly called Indian paintbrush, but that is a different species found to the north. Improbably, Texas paintbrush is in the same plant family as cenizo, which is perhaps the showiest shrub in the region.

Figure 113. Texas paintbrush—March

Scarlet Pea
Indigofera miniata Ortega

Scarlet pea has bilateral petals, one large petal in front and two smaller petals in back (the wide, roundish leaves in the photo belong to American snoutbean). Its flowers are pinkish red, and its leaves are linear to elliptic. As can be seen in relation to the snoutbean leaves (see fig. 36) in the photo, scarlet pea flowers are very small. This low-growing, native legume is common throughout the Sand Sheet. It produces seeds eaten by northern bobwhites and is good to excellent forage for wildlife and livestock.

Figure 114. Scarlet pea—April

Cardinal Feather
Acalypha radians Torr.

It would not be apparent from appearance to the nonbotanist that these were the same species. That is because one is male and the other female, with obviously different appearances. The male version has many-flowered red rods, while the female version has small red spikes. Both have roundish, fuzzy, deeply lobed leaves and both are relatively low growing, up to about one foot high. This species is one of the most common native forbs of the Sand Sheet, often occupying tighter soils and interspaces between native bunchgrasses on sandy soils.

Figure 115. Cardinal feather (male)—April

Figure 116. Cardinal feather (female)—October

Woolly or South Texas Globemallow
Sphaeralcea lindheimeri A. Gray

Figure 117. Woolly globemallow—April

Palm Leaf Globemallow
Sphaeralcea pedatifida (A. Gray) A. Gray

These globemallows have flowers of a similar incandescent orange with yellow centers, but woolly globemallow petals overlap and form a cup, while palm leaf globemallow has five separate petals. The leaves are also different; woolly globemallow has heart-shaped, slightly lobed leaves, while palm leaf globemallow has deeply lobed leaves that look almost like they are branched.

Woolly globemallow is an endemic native on very sandy soils in the Sand Sheet. Palm leaf globemallow prefers tighter soils. White-tailed deer and cattle eat its leaves.

Figure 118. Palm leaf globemallow—April

Turks's Cap

Malvaviscus arboreus Dill. ex Cav. var. drummondii (Torr. & A. Gray) Schery

Turk's cap has a single, partially open red flower with spectacular red stamens projecting about twice as far from the petals as the petals are deep. Its leaves are basically round with points at the end and are conspicuously veined. It is common in shady areas in the eastern part of the Sand Sheet, especially within oak mottes. Deer eat the leaves and flowers.

Figure 119. Turk's cap—May

Weeds and Forbs
Inconspicuously Flowered Forbs

CROTONS

According to one of the Botanical Guides, fourteen species of crotons can be found in South Texas. We believe the five pictured in figures 120–129 are the only ones commonly found in the Sand Sheet. (Texas croton, whose Spanish name is *tinajera*, is not common in the Sand Sheet, although it is sometimes found in the northeastern reaches.)

Woolly Croton
Croton capitatus Michx.

Woolly croton is one of the most common herbaceous plants of the Sand Sheet. It has pubescent, broadly elliptic or narrowly ovate leaves. It is distinguishable from the other four crotons by the shape and pubescence of its leaves, and, in spring, its tall seedhead cones. In the fall, its seeds have a reddish dot in the center. It is often referred to as *tinajera* or "doveweed."

Woolly croton seeds are one of the most important foods of northern bobwhites, mourning doves, and Rio Grande wild turkeys. Woolly croton responds well after soil disturbance or overgrazing. It is easily one of the most important wildlife plants of the region.

Figure 120.
Woolly croton
bunch—October

Figure 121. Woolly croton—May

Figure 122. Woolly croton seedheads—September

Northern Croton (*Vente Conmigo*)
Croton glandulosus L.

Northern croton grows low to the ground, and its seed stalks often appear to spread in a lattice-like arrangement. In early spring, the seed stalks have a reddish tinge. Its leaves are elliptic or oblong and have toothed margins.

Northern croton is common on sandy loam and coarse sandy soils. It is often found growing side by side with woolly croton. It is an excellent seed producer and provides a favored food for northern bobwhites, mourning doves, and Rio Grande wild turkeys.

Figure 123. Northern croton—May

Figure 124. Northern croton seedheads—September

Cory's Croton
Croton coryi Croizat

Cory's croton is easily recognizable by the striking silvery-white appearance of its leaves and its growth in large bunches. Its leaves are ovate and fuzzy, and its seedheads are tall, cone-like, and multi-flowered. Cory's croton is an endemic found on very sandy soils. The seeds are eaten by game birds.

Figure 125. Cory's croton —October

Figure 126. Cory's croton seedheads— September

Parks' Croton
Croton parksii Croizat

Parks' croton is the tallest of the five crotons in this book. It has long, oblong leaves that are slightly wider toward the base than toward the end. It is endemic to Texas and an important food plant for mourning doves and northern bobwhites on very sandy soils. Parks' croton is often common in dune fields and disturbed sites.

Figure 127. Parks' croton—October

Figure 128. Parks' croton—April

Prairie Tea

Croton monanthogynus Michx.

Prairie tea is a common annual of disturbed sites on heavier soils. It can be distinguished from northern croton by the smooth edges to the leaves. Overall the plant has a blue-green color. The seeds are relished by mourning doves and northern bobwhites.

Figure 129. Prairie tea—June

LOW-GROWING FORBS

Hoary Milkpea
Galactia canescens Benth.

Hoary milkpea is a low-growing plant with veined, rounded or broadly ovate leaves. American snoutbean has similar-looking leaves (see figs. 35 and 36) and is often without its yellow flowers, which makes it easy to confuse with hoary milkpea. Hoary milkpea's leaves, however, are narrower than those of American snoutbean and are less prominently veined. It is good forage for livestock and wildlife, and an important food source for northern bobwhites on very sandy soils. It is often a common understory plant in native grasslands.

Figure 130. Hoary milkpea—March

Figure 131. Hoary milkpea—October

Beach Groundcherry

Physalis cinerascens (Dunal) Hitchc. var. spathulifolia (Torr.) J. R. Sullivan

Beach groundcherry is a shade-loving plant with elliptic leaves whose margins are slightly toothed and curl inward. Perhaps its most distinctive characteristic, at least among the low-growing plants in this book, are its flowers, which droop downward. The corollas are yellow with purple centers. Beach groundcherry is a native annual that produces fruits and seeds consumed by many wildlife species. Deer eat the foliage.

Figure 132. Beach groundcherry—March

Ivy Leaf Ground Cherry
Physalis hederifolia A. Gray

Ivy leaf ground cherry has ovate to rounded leaves with irregular margins. Its leaves are the main visual key to tell it apart from beach groundcherry (compare figs. 132 and 133). Later in the spring it will produce yellow flowers with purple centers. It is a shade-loving native annual that produces fruits and seeds consumed by many wildlife species. Deer also consume the foliage.

Figure 133. Ivy leaf ground cherry—April

Prostrate Sandmat
Chamaesyce prostrata (Aiton) Small

Prostrate sandmat is appropriately named, because it is both prostrate and forms a mat. In winter it is reddish, as shown in figure 134. It tends to be more common in drought years.

It is a diminutive native species that occupies harsh sites and disturbed areas. White-tailed deer frequently consume the leaves, and the seeds are eaten by game birds.

Figure 134. Prostrate sandmat—November

Pigeon Berry
Rivina humilis L.

Pigeon berry's most obvious visual clue is its clusters of red berries, and it is usually found in the shade at the base of a tree. Its leaves are ovate/lanceolate with toothed margins. It is a good wildlife plant that produces seeds eaten by many bird species. It is also a prolific bloomer and a good pollinator plant.

Figure 135. Pigeon berry—November

PLANTAINS AND DALEAS

Plantains and daleas share a visual first impression: the tall cones of tiny flowers at the ends of the stalks bring to mind bottlebrushes for some observers. For that reason, they have been grouped together here. The daleas, however, have one characteristic that the plantains don't: their inflorescences all have little upward-pointing, dark red or brown hairs.

Redseed Plantain

Plantago rhodosperma Decne.

Redseed plantain's inflorescences have the appearance of green rods with small cream-colored or red flowers interspersed with small green leaves, called "bracts," that are about the same size as the flowers (see fig. 136, where the flowers are just beginning to turn red). Its leaves are broad, wider toward their ends, and rounded (obovate) or pointed (oblanceolate). They grow in a low clump at the plant's base, which may be the most obvious way to differentiate between redseed and Hooker's plantain (compare figs. 136 and 138). It is a prolific winter annual that provides forage for livestock and wildlife. The seeds are eaten by northern bobwhites.

Figure 137. Red-seed plantain seedheads—April

Figure 136. Redseed plantain—April

Hooker's Plantain
Plantago hookeriana Fisch. & C. A. Mey.

Hooker's plantain closely resembles redseed plantain, but its inflorescences are thicker and have small white flowers and a fuzzy appearance. Its linear leaves form a rosette at the base and are not as prominent as those of redseed plantain (compare figs. 136 and 138). It is a prolific winter annual that provides forage for livestock and wildlife, and the seeds are eaten by northern bobwhites.

Figure 138. Hooker's plantain bunch—March

Figure 139. Hooker's plantain seedheads—March

Golden Dalea
Dalea aurea Nutt. ex Pursh

Golden dalea is also a "bottlebrush" plant, and its leaves help distinguish this species from the other daleas. They are "compound," meaning that several leaves grow from the same point on the stem and are elliptic with white-tinged margins. They are grouped in small clusters at widely separated intervals on the stem, and they are pubescent, or fuzzy. The flowers are yellow.

It is common on sandy loam and very sandy soils of the Sand Sheet and is good forage for livestock and wildlife.

Figure 140. Golden dalea—May

Dwarf Dalea
Dalea nana Torr. ex A. Gray

As the name suggests, dwarf dalea is a very small plant. Its flowers are yellow, three-lobed, and bilaterally symmetrical. Its leaves are elliptic. Both deer and cattle eat the plant.

Figure 141. Dwarf dalea—April

Pussyfoot Dalea
Dalea obovata (Torr. & A. Gray) Shinners

Pussyfoot dalea is endemic to Texas. Its flowers are greenish white, with upturned red hairs. Its leaves are opposite, broadly elliptic, and pubescent. Its inflorescences give an overall impression of bottlebrushes. It is common on very sandy soils.

Figure 142. Pussy-foot dalea—May

Figure 143. Pussyfoot dalea seedhead—April

Wedgeleaf Prairie Clover
Dalea emarginata (Torr. & A. Gray) Shinners

Wedgeleaf prairie clover is an annual legume about two feet high. Cone-like flower heads ringed with small purple blooms branch from single stems. It has a strong citrus odor when crushed. It is eaten by deer and livestock and is usually found on very sandy soils.

Figure 144. Wedgeleaf prairie clover—May

Prairie Mexican Clover
Richardia tricocca (Torr. & A. Gray) Standl.

Prairie Mexican clover has groups of four pointy, lanceolate leaves that radiate outward at about ninety-degree intervals or more. The small white flowers produce seeds eaten by northern bobwhites and mourning doves. It is common on sandy soils.

Figure 145. Prairie Mexican clover—April

Prostrate Bundleflower
Desmanthus virgatus (L.) Willd.

Prostrate bundleflower is a perennial found on heavier soils. It is a low-growing viny legume with cream-colored flowers. Later it has tan legume pods with brown seeds. It is good forage for livestock and wildlife, and its seeds are eaten by northern bobwhites. A commercial seed source called Balli Germplasm is available for planting.

Figure 146. Prostrate bundleflower—May

Prairie Acacia
Acaciella angustissima (Mill.) Britton & Rose

Prairie acacia is a colony-forming, herbaceous legume without thorns. The leaves and flowers are similar to those of prostrate bundleflower but grow upright. The seedpod is smooth, papery, and reddish brown. The stems are often reddish, turning slate gray at maturity. It is found on red sandy loam soils of the Sand Sheet, typically in areas protected from heavy grazing or mechanical disturbance, often in road rights-of-way, old cemeteries, or very high-quality native ranges. Prairie acacia is relished by white-tailed deer and cattle, and the seeds are eaten by northern bobwhites. A commercial seed source called Rio Grande Germplasm is available for restoration plantings.

Figure 147. Prairie acacia—August

Figure 148. Prairie acacia—August

TALLER FORBS AND WEEDS

Queen's Delight
Stillingia sylvatica L.

Queen's delight is frequently encountered in the Sand Sheet. Its oblanceolate leaves curve upward. Its flower cluster is a prominent spike (see fig. 149) that by late spring or early summer will have roundish green lobes at the base that contain seeds (see fig. 150). Cattle generally avoid grazing the plant, but it is an important food plant of northern bobwhites, white-winged doves, and mourning doves.

Figure 149. Queen's delight—March

Figure 150. Queen's delight seedhead—May

Western Ragweed
Ambrosia psilostachya DC.

In the spring, western ragweed can be identified by its leaves, which are ovate-lanceolate with multiple deep, prominent lobes. (Note to the nonbotanist: even though the lobes tend to dominate the shape, in the case of "deeply lobed" leaves, one must envision what the leaves would look like without lobes at all to come up with a leaf shape.)

Western ragweed is another Sand Sheet plant that changes appearance from season to season, as illustrated by these photos. In the summer, western ragweed is leafy and does not have seed columns (see fig. 151). In the fall, it has tall, multiflowered seedheads (see fig. 152) that, together with its smell and leaf shape, make it easy to identify.

Western ragweed is a very important food source for northern bobwhites in the Sand Sheet. It is poor forage and can be a prolific weed in pastures following rainfall cycles.

Figure 152. Western ragweed—November

Figure 151. Western ragweed—May

Woolly Tidestroma
Tidestromia lanuginosa (Nutt.) Standl.

This plant looks something like a big northern croton in the fall, with reddish stems and small white flowers, although it grows in much bigger clumps than northern croton. In the winter, however, it comes loose from its roots and becomes a tumbleweed. It can be an important cover plant on some soils. The seeds are eaten by northern bobwhites, and deer occasionally browse it.

Figure 153. Woolly tidestroma—November

Figure 154. Woolly tidestroma—January

Southern Pepperweed
Lepidium austrinum Small

Southern pepperweed is another plant with tall flower stalks. It has tiny flowers on small stems radiating out from the stalk in all directions, one flower to a stem. This sparsely flowered, delicate-appearing stem construction is its most prominent visual clue. Its leaves are linear, with some slightly toothed, and sharp pointed. It is a common winter annual in the Sand Sheet. It appears with green as well as red inflorescences. Deer and livestock eat the foliage, and northern bobwhites consume the seeds.

Figure 155. Southern pepperweed—March

Figure 156. Southern pepperweed—April

Blackroot
Pterocaulon virgatum (L.) DC.

Blackroot is a common perennial on tighter soils in the Sand Sheet. Its stalks are tall in relation to the size of the bunch and radiate upward and outward from a narrow base. It produces small purple flowers in the summer. It is easy to identify by its linear leaves, which are dark green on one side and white and fuzzy on the other. It is not typically grazed by cattle but is often an attractive pollinator plant.

Figure 160. Blackroot—May

Texas Vervain
Verbena halei Small

Texas vervain is a tall plant with branched seed stems and small purple flowers extending several inches along the stems. It is a good pollinator plant.

Figure 161. Texas vervain—March

Flatsedge
Cyperus retroflexus Buckley

Flatsedge is neither a grass nor a forb, and that is why it has been placed at the end of this section. Its most recognizable feature is the cluster of yellow spikes that form the heads. It also has distinctively long, slender, lanceolate leaves in groups of three. Wildlife and cattle frequently consume the foliage. (Several species of sedges are common on many soils in the Sand Sheet; identification of specific species is difficult in the field. Some species are common in dry uplands, whereas others are found in wet areas exclusively.)

Figure 162. Flatsedge—April

Figure 163. Flatsedge— March

Grasses

BLUESTEMS

Big Bluestem
Andropogon gerardii Vitman

Big bluestem is one of the tallest grasses (the bunch in fig. 164 is over six feet tall), and it is found only on protected or well-managed sites. It is more common in the eastern part of the Sand Sheet, although it is relatively rare there. It tends to grow in isolated clumps but occasionally grows in widespread colonies. Its leaves are positioned more or less evenly all the way up the stem and tend to face a single direction. At the tips of the seed stalks are three or four individual seedheads that spread out in a "turkey foot" fashion. Big bluestem is excellent forage for livestock and a good cover plant for wildlife.

Figure 166 shows a rare sight: the flowering of big bluestem, which lasts for only a few hours every year before the blooms fall off.

Figure 165. Big bluestem seedheads—October

Figure 164. Big bluestem clump—October

Figure 166. Big bluestem seedheads—August

Little Bluestem
Schizachyrium scoparium (Michx.) Nash

Little bluestem and seacoast bluestem are difficult to tell apart and are very closely related. Little bluestem usually grows in distinct clumps, whereas seacoast bluestem usually grows in colonies; but they are often found growing together. In the spring and summer, before the seedheads have sprouted, the best visual key is the color of the leaves. Little bluestem has some reddish purple to its leaves, while seacoast bluestem is more of a blue green (compare figs. 168 and 171).

Figure 167. Little bluestem—November

Figure 168. Little bluestem—summer

Figure 169. Little bluestem seedheads—November

Seacoast Bluestem
Schizachyrium littorale (Nash) E. P. Bicknell

In the winter it is easier to tell little bluestem and seacoast bluestem apart because the seedheads are slightly different. Seacoast seedheads are fuzzy white and appear larger as a result. Little bluestem seedheads are less robust, but also white and similar looking upon examination, just less prominent. In the winter, seacoast bluestem colonies can resemble snowfall from a distance (see fig. 173).

Both seacoast and little bluestem are good to excellent forage and are the dominant grasses on well-managed pastures in the Sand Sheet. Both species are important as nesting cover to northern bobwhites. Seacoast bluestem is more common in the Sand Sheet. Little bluestem is found on tighter soils, whereas seacoast bluestem is common on very sandy sites. An adapted seed source of little bluestem is available for reseeding efforts, sold as Carrizo Blend.

▶ Figure 170. Seacoast bluestem—November

▶ Figure 171. Seacoast bluestem—summer

▶ Figure 173. Seacoast bluestem—December

▲ Figure 172. Seacoast bluestem—October

Tanglehead

Heteropogon contortus (L.) P. Beauv. ex Roem. & Schult.

Tanglehead grows in thick bunches, about the same height as seacoast and little bluestem (around two and a half to four feet). Its chief visual characteristic is the long, dark brown or black seed tips (fig. 174), which tangle with each other in the winter (fig. 175) and give this grass its name. Tanglehead is a member of the bluestem group.

Tanglehead has expanded significantly in the last thirty years and now threatens to crowd out other grasses in some areas. The cause is not known, but some experts believe it has to do with the reduction of grazing pressure on South Texas pastures over that time. It forms colonies that expand rapidly and leave little space between stalks, making it difficult for quail and other birds to use. Cattle seem to avoid eating it when other options are available, and burning it just seems to promote its expansion, unless the burned area is grazed immediately after the fire.

Figure 174. Tanglehead clumps—September

Figure 175. Tanglehead seedheads—November

Crinkleawn
Trachypogon spicatus (L. f.) Kuntze

Crinkleawn is easy to identify, except in winter when it has lost its color and its seed stems have dropped off. In the spring and summer, crinkleawn's blades are noticeably blue green, and they tend to bend from the bunch at sharp angles. The flowers are bright yellow and are encapsulated in the seed stems (see fig. 178). The "awns," or hairlike appendages that come from the seed, are crinkly and twisted (see fig. 179), hence the name "crinkleawn." It is good to excellent forage for livestock, and a good cover plant for wildlife. Crinkleawn is common in well-managed pastures but declines with overgrazing.

Figure 177. Crinkleawn—summer

▲ Figure 176. Crinkleawn colony— November

◄ Figure 178. Crinkleawn seedheads— October

▶ Figure 179. Crinkleawn seedhead with awns— November

Yellow Indiangrass
Sorghastrum nutans (L.) Nash

Yellow Indiangrass leaves appear blue green most of year, like those of crinkleawn, but they are wider and do not bend off from the stalk at near right angles, as crinkleawn leaf blades do. The seed stalks are distinctively golden and tall, often chest high. It has a single seedhead on each stalk, which is not encapsulated like those of crinkleawn, and its seeds have short awns. It is excellent forage for livestock and a good cover plant for wildlife. It is usually found on the least-grazed sites or very well-managed pastures, often within or adjacent to stands of switchgrass, crinkleawn, seacoast bluestem, and big bluestem.

Figure 180. Yellow Indiangrass—November

Figure 181. Yellow Indiangrass seedheads—August

Bushy Bluestem
Andropogon glomeratus (Walter) Britton, Sterns & Poggenb.

Bushy bluestem is a tall, stemmy bunchgrass that is often found in wet areas and has a dense cluster of hairy seeds at the ends of its stalks. Given its seedhead structure, it is hard to miss. It is a bunchgrass with poor grazing value and is generally one of the last plants eaten by livestock.

Figure 182. Bushy bluestem—November

Figure 183. Bushy bluestem seedheads—October

Silver Bluestem
Bothriochloa laguroides (DC.) Herter

Figure 184. Silver bluestem—June

Figure 185. Silver bluestem seedhead—May

Cane Bluestem
Bothriochloa barbinodis (Lag.) Herter

Silver bluestem leaves are true green, often with purple or red margins. Its stems are long and smooth. Its cream-colored seeds are clustered at the ends of the seed stalks. Cane bluestem is similar in appearance; both have fanlike, bunchy white seedheads, but those of cane bluestem are more widely branched. Cane bluestem is generally found on sandier, dryer soils. True differentiation of cane and silver bluestems is difficult, and plants often possess characteristics of both species.

Silver bluestem is good to fair forage and is common in much of the Sand Sheet, especially on tighter soils, in clayey depressions, and along drainages. An adapted seed source of longspike silver bluestem, a similar species, is sold as Starr Germplasm. Cane bluestem is fair to good forage for livestock and is usually found on reddish sandy loam soils.

Figure 186. Cane bluestem—September

Figure 187. Cane bluestem seedhead —September

Kleberg Bluestem
Dichanthium annulatum (Forssk.) Stapf

Kleberg bluestem is an African grass introduced to South Texas in the early 1900s, and it was seeded following brush work and for erosion control in many areas. It has invaded other areas since and is commonly found near ranch roads, along rights-of-way, and in areas with vehicular traffic. Today, this plant represents one of the most extensive threats to native plant diversity in South Texas. Areas dominated by Kleberg bluestem are poor habitat for most wildlife. Forage value varies throughout the year. Control of Kleberg bluestem is difficult, and it is extremely invasive following soil disturbance, fire, or overgrazing.

Figure 188. Kleberg bluestem —March

Figure 189. Kleberg bluestem seedhead—March

Figure 190. Kleberg bluestem stalk—March (note ring of white hairs)

King Ranch Bluestem
Bothriochloa ischaemum (L.) Keng

King Ranch bluestem is similar in appearance to Kleberg bluestem but is less common in the Sand Sheet. It is usually found near roads, particularly caliche roads. Probably the most reliable way to tell them apart is to look at the stems. Kleberg bluestems stalks have a small ring of white hairs on each stem node (see fig. 190). King Ranch bluestem stalks do not. Kleberg seedheads are deep maroon and usually more robust than King Ranch bluestem seedheads. Both turn yellow to straw colored in fall and winter.

Be careful not to confuse these bluestems with multiflower false Rhodes grass, whose seedheads are similar. Multiflower false Rhodes grass is taller and has broad leaves instead of narrow blade-shaped leaves, and the seedheads lack maroon coloration at maturity (see figs. 251 and 252).

King Ranch bluestem was introduced for forage provision and soil stabilization and was seeded extensively in the last century. It provides fair to good forage depending on the growing site and management. Areas dominated by King Ranch bluestem are poorer habitat for most wildlife than sites dominated by native grass. Kleberg bluestem is much more common in the Sand Sheet than King Ranch bluestem. Neither species grows well in very sandy soils.

Figure 191. King Ranch bluestem—September

Figure 192. King Ranch bluestem seedheads—July

PANICUMS

Switchgrass
Panicum virgatum L.

Switchgrass can grow up to head high and is blue-green to green in the growing season. The seedheads are openly branched and have many seeds. The presence of switchgrass is an indicator of good pasture conditions. It is excellent forage for livestock, and a good seed producer and cover plant for wildlife. Switchgrass is usually found in areas also supporting big bluestem, yellow Indiangrass, crinkleawn, and seacoast bluestem.

Figure 193. Switch-grass clump —November

Figure 194. Switch-grass seedheads —November

Guineagrass
Urochloa maxima (Jacq.) R. Webster

Guineagrass has long, curled leaves that grow in tight, luxuriant green bunches, usually in shaded areas, and can reach heights of four to six feet. It will sometimes vine into brush canopies (see fig. 195). Guineagrass is invasive following fire, overgrazing, drought, or other disturbances. It is a nonnative grass from Africa that is thought to be spread by birds. Guineagrass was uncommon in the Sand Sheet and South Texas just a few decades ago, but today it has invaded most of the region. It is excellent forage and the seeds are eaten by many birds, but it can outgrow grazers in periods of copious rainfall. It often grows into the canopies of brush mottes, making them unnaturally susceptible to fire. Guineagrass invasion into native grasslands and the subsequent loss of plant diversity is a major conservation concern for the habitats of the Sand Sheet.

Figure 195. Guineagrass clump—August

Figure 196. Guineagrass seedheads —September

Texas Panicum or Texas Signalgrass
Urochloa texana (Buckley) R. Webster

Texas panicum is short in stature and has long, spear-shaped, erect leaves that often extend above its seedheads (see fig. 197). Its seedheads are closely bunched and columnar (see fig. 198). This annual grass is usually found in shady areas and is commonly known as "conchograss" by ranchers. Texas panicum is a native annual grass and is one of the most important food plants for mourning doves and northern bobwhites, especially in disturbed areas, croplands, and very sandy soils in the Sand Sheet. It is excellent forage for livestock and a good seed producer for wildlife.

Figure 197. Texas panicum bunch—August

Figure 198. Texas panicum seed-head—August

Fall Witchgrass
Digitaria cognata (Schult.) Pilg.

Fall witchgrass is a low-growing grass, one foot high or less, that has many-branched seedheads, with one seed at the tip of each branch of the stem. That distinguishes it from southern witchgrass, which is similar in overall appearance and sometimes occupies the same areas but has many seeds per stem. A visual key for fall witchgrass is that its leaves are typically serrated on one side only (see fig. 200). It is common but inconspicuous over much of the Sand Sheet. Fall witchgrass is fair to poor forage on most sites.

Figure 199. Fall witchgrass bunch—April

Figure 200. Fall witchgrass seedheads—May

Southern Witchgrass
Panicum capillarioides Vasey

Southern witchgrass is a small bunchgrass with many seedheads emanating all the way from the base, leaving the visual impression of a sprawling, every-which-way gossamer of delicate stems and seeds. It is easy to confuse with fall witchgrass, but fall witchgrass has only one seed per stem, while southern witchgrass has many. It is common on sites where the soil was recently disturbed, or where grazing has been intense. It provides fair to good forage for livestock, and the seeds are consumed by northern bobwhites and mourning doves.

Figure 201. Southern witchgrass bunch—April

Figure 202. Southern witchgrass seedheads—May

BRISTLEGRASSES
Knot Grass

Setaria reverchonii (Vasey) Pilg. ssp. firmula (Hitchc. & Chase) W. E. Fox

As with all bristlegrasses, the best indicator of the species is the seedheads. Knot grass has short, broad leaves and long, thin seedheads with irregular seed density. Knot grass seeds grow on the main stem and more or less wrap upward around the seed stems. Knot grass is commonly found on very sandy sites. It is good forage for livestock, and a copious producer of seeds eaten by game birds.

Figure 203. Knot grass bunch—March

Figure 204. Knot grassseedheads—March

Plains Bristlegrass
Setaria leucopila (Scribn. & Merr.) K. Schum.

Plains bristlegrass has broad, spear-shaped leaves, like the other bristlegrasses. Its seedheads are long, fuzzy, and usually drooping. They are somewhat similar in appearance to Texas bristlegrass seedheads, but plains bristlegrass grows in thicker and taller bunches (see fig. 205). Plains bristlegrass is an important native grass in tighter soils and shaded areas. It is good to excellent forage for livestock, and a good producer of seeds eaten by northern bobwhites. A commercial seed source called Catarina Blend is available for rangeland seedings.

Figure 205. Plains bristle-grass bunch—July

Figure 206. Plains bristlegrass seedheads—June

Knotroot Bristlegrass
Setaria parviflora (Poir.) Kerguélen

Knotroot bristlegrass seedheads are different in appearance from other bristlegrass seedheads, as can be seen in figure 208 versus figures 206 and 210. In addition, knotroot bristlegrass seedheads are orange. Knotroot bristlegrass is common in very sandy soils and wetland areas. It is in general poor forage but is a good seed producer for game birds.

Figure 207. Knotroot bristlegrass bunch—May

Figure 208. Knotroot bristlegrass seedhead—May

Texas Bristlegrass

Setaria texana W. H. P. Emery

Texas bristlegrass is a shade-loving, diminutive species usually found under the canopy of brush mottes. Its seedheads bear a close resemblance to knot grass seedheads (see fig. 204). Texas bristlegrass seedheads, however, grow on short stems that parallel the seed stems, rather than on the seed stem itself. Otherwise, it is similar in appearance to plains bristlegrass, except it is never taller than about two feet, and its seedheads, while bristly, are narrower than those of plains bristlegrass and much less bristly. Texas bristlegrass is good forage found in brush canopies.

Figure 209. Texas bristlegrass bunch—May

Figure 210. Texas bristlegrass seedheads—May

PASPALUMS

Thin Paspalum
Paspalum setaceum (Michx.)

Thin paspalum is a common grass in the Sand Sheet that has short, pubescent (hairy), lanceolate (spear shaped) leaves. Its seedheads are distinctive, forming elongated clusters at the end of the stem. Its seeds are round but flat on one side, an important clue. Thin paspalum is fair to good forage for livestock and produces seeds that are frequently eaten by northern bobwhites.

Figure 211. Thin paspalum bunch—April

Figure 212. Thin paspalum seedheads—March

Brownseed Paspalum
Paspalum plicatulum Michx.

Brownseed paspalum seedheads are distinctive in appearance, forming brown-seeded clusters growing at about forty-five-degree angles from the stem. Brownseed paspalum is a bunchgrass found in wet areas and well-managed sandy pastures. It is good to excellent livestock forage and an indicator of good range conditions, as well as a good nesting substrate and food source for northern bobwhites.

Figure 213. Brownseed paspalum bunch—May

Figure 214. Brownseed paspalum seedheads—December

Gulfdune Paspalum
Paspalum monostachyum Vasey

Gulfdune paspalum is a common but obscure grass that often grows in the same bunches with other grasses, making it hard to find. Its leaves coil in on themselves, forming tubes. It is a native grass found on very sandy soils, including on dune fields, depressions, and sandy prairies, often with seacoast bluestem. It is poor forage, but an important grass for soil stabilization over much of the Sand Sheet. When it does produce seeds, they are likely consumed by northern bobwhites and Rio Grande wild turkeys.

Figure 215.
Gulfdune
paspalum—May

SIGNALGRASSES

Fringed Signalgrass
Urochloa ciliatissima (Buckley) R. Webster

Fringed signalgrass looks somewhat like knot grass (see fig. 204). Both have relatively short, broad, spear-shaped leaves. However, signalgrass seedheads are noticeably fuzzy, while knot grass seedheads are clean. Also, fringed signalgrass leaves are shorter, wider, and pubescent. Fringed signalgrass is one of the most common grasses of sandy soils in the region, forming dense ground cover. Several other signalgrasses are common in the Sand Sheet but are hard to differentiate in the field. Most have value similar to that of fringed signalgrass.

Figure 216. Fringed signalgrass bunch—March

Figure 217. Fringed signalgrass seedheads—March

Arizona Cottontop
Digitaria californica (Benth.) Henr.

Arizona cottontop has hairy leaves, but they tend to be blue green; fringed signalgrass and Arizona cottontop do not otherwise resemble each other. Arizona cottontop is found mainly on tight sandy loam soils and saline areas of the Sand Sheet. It is an upright bunchgrass and is good to excellent forage for livestock.

Figure 218. Arizona cottontop bunch—June

Figure 219. Arizona cottontop seedheads—April

THREEAWNS

Purple Threeawn
Aristida purpurea Nutt. var. purpurea

Purple threeawn is a drapey bunchgrass about two feet high. Figure 219 illustrates why this grass is called "purple threeawn." It is not always purple, however (see fig. 221). It is generally regarded as poor forage for livestock.

Figure 220. Purple threeawn—March

Figure 221. Purple threeawn —December

Blue Threeawn
Aristida purpurea Nutt. var. nealleyi (Vasey) Allred

Blue threeawn is taller than purple threeawn (two to four feet, about twice the height of purple threeawn), and its awns (the hairs that issue from the seed) are longer, up to an inch in length.

Both purple threeawn and blue threeawn have good value as cover to wildlife and are used for nesting by northern bobwhites when other, more palatable grasses are eliminated by grazing. Threeawns are natural colonizers of disturbed sites and often grow in shallow soils and dry areas that support no other grasses. Both blue and purple threeawn are common in the Sand Sheet.

Figure 222. Blue threeawn colony—May

Figure 223. Blue threeawn seedheads—November

LOVEGRASSES

Lehmann Lovegrass
Eragrostis lehmanniana Nees

Lehmann lovegrass is an indistinct and very common inhabitant of the Sand Sheet. It takes on different appearances at different times of the year. In the spring, it is bright green, upright, and delicate looking (see fig. 224). In wet, windy weather, it blows over into a gray-white, wispy carpet, as seen in figure 225. It is an African grass introduced to South Texas and used for forage provision and erosion control. It is poor forage except at green-up in early spring or following rainfall.

Lehmann lovegrass mats formed by copious growth limit the growth of other plants and create poor habitat for wildlife. This grass responds with aggressive, invasive growth following fire or drought.

Figure 224. Lehmann lovegrass— spring

Figure 225. Lehmann lovegrass— November

Plains Lovegrass
Eragrostis intermedia Hitchc.

Plains lovegrass closely resembles Lehmann lovegrass. One obvious difference is that Lehmann lovegrass grows in colonies, whereas plains lovegrass is a bunchgrass. Plains lovegrass is generally fair forage and is a minor component of Sand Sheet grasslands. Full disclosure: you can't reliably tell the difference between these and several other Sand Sheet lovegrasses at certain times of the year without seedheads and botanical measurements.

Figure 226. Plains lovegrass—July

Red Lovegrass
Eragrostis secundiflora J. Presl

Red lovegrass is one of the easy lovegrasses to identify, because of its pinkish to reddish color. It is a small bunchgrass and its leaves are strongly blue green, which makes it easy at first to confuse with crinkleawn in the spring, if there are no seedheads present. Red lovegrass grows in small bunches, whereas crinkleawn grows in colonies (see fig. 176). From a distance, it can also resemble natal grass (see fig. 258). Natal grass, however, is much more wispy and delicate and has taller seedheads.

Red lovegrass is a common and widespread native grass of the Sand Sheet. It is generally considered poor forage, and it can be dominant on overgrazed pastures. Northern bobwhites and turkeys consume the seeds. A commercial seed source of this grass, marketed as Duval Germplasm, is useful in restoring disturbed sites in the Sand Sheet.

Figure 227. Red lovegrass bunch—April

Figure 228. Red lovegrass seedheads—July

Wilman Lovegrass
Eragrostis superba Peyr.

Wilman lovegrass is a tall, stemmy bunchgrass with light-colored and distinctively shaped seedheads up and down the stem that look something like wheat.

This is another grass introduced from Africa. It was historically used in many USDA seeding projects to provide forage, and as a fast cover plant in range seedings. It has fair to poor grazing value, and dense stands can limit use of an area by wildlife because of the resulting negative impact on plant diversity. Wilman lovegrass varies from benign to extremely invasive in parts of the Sand Sheet, depending largely on the soil on which it is found. It is likely to be used by northern bobwhites and turkeys for nesting.

Figure 229. Wilman lovegrass bunch—November

Figure 230. Wilman lovegrass seedheads—March

Gummy Lovegrass
Eragrostis curtipedicellata Buckley

Gummy lovegrass is a small native bunchgrass that is easy to mistake for southern witchgrass (see figs. 201 and 202). A good visual key is that southern witchgrass seed stems lie at almost right angles to the main stem and are sparse, whereas gummy lovegrass stems are profuse. Also, gummy lovegrass seedheads are bunched in a radiating cluster, whereas southern witchgrass seedheads have a single spikelet at the end of the stem. Gummy lovegrass also feels gummy to the touch—hence the name.

Gummy lovegrass is a minor component of native vegetation communities on sandy loam soils and is generally regarded as fair to poor forage. The sticky leaves likely make it unattractive to grazers. The seeds are similar to other types and sizes of seeds known to be consumed by northern bobwhites and Rio Grande wild turkeys.

Figure 231. Gummy lovegrass bunch—March

Figure 232. Gummy lovegrass seedheads—March

Tumble Lovegrass
Eragrostis sessilispica Buckley

Tumble lovegrass is a common, inconspicuous grass and is generally identifiable only in the spring, when the long, sparsely seeded, reddish seedheads are present. Its seed stems also lie at almost right angles to the main stem. It rolls and mounds within other vegetation and eventually detaches and tumbles in great numbers in some years.

Tumble lovegrass is a minor species of grasslands on tighter soils. The foliage is unassuming, but the large, tumbling seedheads can be conspicuous in summer. It is generally regarded as poor to fair forage and is common on hard-grazed sites.

Figure 233. Tumble lovegrass bunch—March

Figure 234. Tumble lovegrass seedheads—March

GRAMAS
Hairy Grama
Bouteloua hirsuta Lag.

Hairy grama is a common bunchgrass with a distinctive, comb-shaped seedhead. It is native and is found on all types of soils in the Sand Sheet; in some years it can be a dominant species. Hairy grama is good to excellent forage. A commercial seed selection called Chaparral Germplasm is available for restoration seed mixes in the Sand Sheet.

Figure 235.
Hairy grama
bunch—November

Figure 236.
Hairy grama
seedheads—May

Slender Grama
Bouteloua repens (Kunth) Scribn. & Merr.

Slender grama has lime-green leaves and many spiky-looking seedheads. It is an important native grass found on a wide variety of soils, from tight sandy loams to dune sands. Dilley Germplasm is a commercial seed source of this plant that is especially well adapted to use in reseeding projects in the Sand Sheet. Slender grama is fair to good forage.

Figure 237. Slender grama bunch—September

Figure 238. Slender grama seedheads—June

Texas Grama
Bouteloua rigidiseta (Steud.) Hitchc.

Texas grama is similar in appearance to slender grama but has more pronounced and prickly spikes on the seed clusters. It grows throughout the winter and produces seeds in early spring. Texas grama is an important native bunchgrass on harsh sites, although it has poor grazing value. It is more common in regions adjacent to the Sand Sheet and is a minor component of Sand Sheet grasslands on heavier-textured soils. A commercial seed source called Atascosa Germplasm is available for reseeding projects.

Figure 239. Texas grama bunch—June

Figure 240. Texas grama seedheads—June

PAPPUSGRASSES
Whiplash Pappusgrass
Pappophorum vaginatum Buckley

Whiplash pappusgrass is a bunchgrass that has long, slender leaves and a fuzzy, columnar seed stem. Its seedheads look much like those of buffelgrass (see fig. 261) but have smaller seedhead burs and whitish seedheads. It is found on saline sites, near *ramaderos*, and in the eastern Sand Sheet on heavier soils (never on coarse sandy soils). It is an important native species of coastal areas, and in some areas of the Sand Sheet it is a dominant grass. It is fair forage for livestock and provides good cover for wildlife on sites that otherwise have little grass cover. A commercial seed source called Webb Germplasm is available for reseeding projects.

Figure 241. Whiplash pappusgrass bunch—March

Figure 242. Whiplash pappusgrass seedhead—August

Pink Pappusgrass
Pappophorum bicolor Fourn.

Pink pappusgrass also somewhat resembles buffelgrass but has noticeably pink seedheads that turn brownish tan at maturity. Its leaves are long, thin, and spiky, in contrast to buffelgrass leaves, which are wider and shorter relative to the size of the plant. Pink pappusgrass is a dominant native grass on some sites, especially in the adjacent Rio Grande Plain. Pink pappusgrass is generally found on sandy loam, clayey, or tighter soils in the Sand Sheet. It is regarded as good forage for livestock and provides nesting sites for northern bobwhites. A commercial seed source called Maverick Germplasm is available for planting.

Figure 243. Pink pappusgrass bunch—August

Figure 244. Pink pappusgrass seedheads—July

WINDMILLGRASSES

Hooded Windmillgrass
Chloris cucullata Bisch.

Before seeding out, hooded windmillgrass seedheads are closely wrapped and erect (see fig. 245). Once they have seeded out, it is hard to mistake this for any other species. Its multiple seedheads spread out like a windmill. The brown to purple seeds hang downward from the curly stems. In late fall and winter, its seedheads become curled and whitish (see fig. 247). Hooded windmillgrass is good forage for livestock, and the seeds are eaten by many wildlife species. A commercial seed source called Mariah Germplasm is available for Sand Sheet reseeding efforts.

Figure 245. Hooded windmillgrass bunch—September

Figure 246. Hooded windmillgrass seedhead—March

Figure 247. Hooded windmillgrass seedheads—November

Shortspike Windmillgrass

Chloris ×subdolichostachya Müll. Berol. (pro sp.) [cucullata × verticillata]

Shortspike windmillgrass is similar in appearance to hooded wind-millgrass, but its seed stems do not curl in random directions like those of hooded windmillgrass, and its seedheads are spiky and whitish to green. The seedheads are generally thinner than those of hooded windmillgrass, and the individual stems of the seedheads are shorter. Shortspike windmillgrass is good livestock forage, and wildlife eats the seeds. A commercial seed source named Welder Germplasm is available for planting in reseeding projects.

Figure 248. Shortspike windmillgrass—October

Figure 249. Shortspike windmill-grass seedheads—August

OTHER SPECIES

Slim Tridens
Tridens muticus (Torr.) Nash

Slim tridens is a stemmy, bunchy, blue-green to gray grass with ridged leaves, and it is found on sandy loam or tighter soils. The seedheads are snowy in appearance when mature and may have a purple hue before ripening.

One of the major rangeland grasses of Texas, slim tridens is found in limited amounts in the true Sand Sheet, usually on tighter sandy loam soils or on shallow, dry sites overlying caliche subsoils. Slim tridens is a very hardy, drought-tolerant native grass. It provides fair to good forage for livestock. The seeds are eaten by many wildlife species and are favored by harvester ants.

Figure 250. Slim tridens—March

Multiflower False Rhodes Grass
Trichloris pluriflora Fourn.

Multiflower false Rhodes grass is a tall bunchgrass (three to five feet) with broad green leaves. Its seedheads are erect and many branched, with small awned seeds on the branches. It is an important native bunchgrass found on tighter sandy loam soils. The seedheads can be mistaken for those of Kleberg or King Ranch bluestems (see figs. 188 and 192).

Often called "four-flower trichloris," multiflower false Rhodes grass is a very important dominant native grass on many sandy loam and tighter soils in the region. It is excellent forage and a good indicator of proper range management. It declines rapidly with overgrazing but can respond impressively following brush work or copious rainfall. It provides excellent wildlife cover. A commercial seed source is available for restoration efforts, sold as Hidalgo Germplasm.

Figure 251. Multiflower false Rhodes grass bunch—May

Figure 252. Multiflower false Rhodes grass seedheads—March

Sacahuiste or Gulf Cordgrass
Spartina spartinae (Trin.) Merr. ex Hitchc.

Sacahuiste, also known as Gulf cordgrass, is common in the eastern part of the Sand Sheet. It forms thick colonies, sometimes too thick to walk through. It provides good nesting cover for northern bobwhites and habitat for small animals (including rattlesnakes). Areas that grow sacahuiste generally will not support other grasses because of soil salinity and alkalinity. It is palatable to livestock only immediately following burning.

Figure 253. Sacahuiste colony—May

Figure 254. Sacahuiste seedhead—May

Pan American Balsamscale
Elionurus tripsacoides Humb. & Bonpl. ex Willd.

Pan American balsamscale is a very common, stemmy grass usually found on heavily grazed sites. Its seeds form smooth, interwoven pods early in the year. The pods break open in the fall and look "scaly." It is poor forage and is often the only grass cover on overgrazed pastures.

Figure 255. Pan American balsamscale bunch—May

Figure 256. Pan American balsamscale seedheads—September

Bermudagrass
Cynodon dactylon (L.) Pers.

Bermudagrass is a turf-forming introduced grass that has been widely naturalized throughout the Sand Sheet. Its seedheads are windmill-like but upturned, with few branches and very small seeds. Bermudagrass is native to Africa but has been widely planted throughout the Sand Sheet to provide forage for livestock and to prevent erosion. It is most common in tighter soils, along roads, and near water sources, where it has often been planted to protect pond dams. Large areas of former cropland often grow bermudagrass, as do pastures that were sprigged with the grass to provide forage or be harvested for hay. Bermudagrass is good livestock forage but represents very poor habitat for most all wildlife.

Figure 257.
Bermudagrass—
September

Natal Grass
Melinis repens (Willd.) Zizka

Natal grass is common in the Sand Sheet. It is a knee-high, stemmy grass with fluffy seedheads that blow easily in even low winds. It often falls over and forms dense mats after a frost. Its seedheads become pink in the fall and are covered in dense hairs. It is an African grass that was introduced in the early 1900s to South Texas. It is poor forage and can be invasive on many sites. The dense foliage and thatch can choke out other plant life.

Figure 258. Natal grass colony—September

Figure 259. Natal grass seedheads—September

Buffelgrass
Pennisetum ciliare (L.) Link

Buffelgrass is an aggressive, introduced bunchgrass with thick, knotty stems and lime-green leaves. It is similar in appearance to whiplash pappusgrass (figs. 241 and 242) and pink pappusgrass (figs. 243 and 244). It often grows in dense stands that outcompete native plants. Buffelgrass was extensively seeded in South Texas to provide forage for livestock, but it has tended to crowd out native grasses and spread from the areas in which it was planted.

Buffelgrass is also native to Africa. Although it is excellent livestock forage, the invasive nature of the plant, coupled with its dominance over all other herbaceous plants in the region, makes it a poor choice for wildlife habitat. Northern bobwhites are known to nest and roost in buffelgrass, but it otherwise represents poor habitat for them. Control of established stands of buffelgrass or conversion back to native habitat once it is established is difficult.

Figure 260. Buffelgrass colony—September

Figure 261. Buffelgrass seedheads —September

Texasgrass
Vaseyochloa multinervosa (Vasey) Hitch.

Texasgrass is endemic to the Sand Sheet. It is often found in the canopy or shade adjacent to brush or oak mottes. It has blue-green leaves, often with a reddish tinge. Its seedheads are tall and have many clusters of large, heavy seeds, sometimes resulting in the seedheads and stalks lying over at or near maturity.

Texasgrass is good to excellent forage and is generally absent from overgrazed sites. It is a copious seed producer, and the seeds are eaten by northern bobwhites and Rio Grande wild turkeys.

Figure 262. Texasgrass seedheads—March

Sand Dropseed
Sporobolus cryptandrus (Torr.) A. Gray

Sand dropseed is a common native grass found on all Sand Sheet soils, usually in small bunches. There is usually a unique flag leaf at the point of the stem where the seedheads emerge (see fig. 264). Its seedheads have a longitudinal groove that contains the tiny, amber-colored seeds (see fig. 265). Turkeys can strip out the seeds by pulling their beaks along the stems. Its stem nodes are characteristically ringed with a star of hairs.

Sand dropseed is fair to good forage, depending on the site and ecotype of the plant. Two distinct growth forms are common in the Sand Sheet, including a small, stemmy ecotype found on tighter soils, and a robust, leafier type found on very sandy soils as part of the climax vegetation. The seeds are also eaten by northern bobwhites. A commercial seed source called Nueces Germplasm is available for reseeding efforts in the Sand Sheet.

Figure 264. Sand dropseed flag leaf—September

Figure 263. Sand dropseed bunch—September

Figure 265. Sand dropseed seedheads—September

Giant Reed or Carrizo Cane
Arundo donax L.

Giant reed is a tall grass (over six feet) that has curled, lanceolate leaves beginning about halfway up the stalk. The seedheads are conspicuously bushy. This introduced species was planted for erosion control along streams and in active sand dunes. It is difficult to control. Giant reed can provide good cover for wildlife, and livestock will graze it when it is small. It is considered a problematic invasive plant along streams and rivers and is a subject of federal control programs in many areas.

Figure 266. Giant reed—October

Shrubs without Thorns

As even a casual observer knows, there is a very hazy line in the transition from shrubs to trees in South Texas. We have used considerable latitude in grouping these plants for purposes of this book. We have divided shrubs into two groups: thorned and unthorned.

Purple Sage (*Cenizo*)
Leucophyllum frutescens (Berl.) I. M. Johnst.

Purple sage is unmistakable after rains in the spring (although it can bloom throughout the year). Its lavish adornment of violet and pink flowers make it so. Its leaves are obovate, silvery gray green, and fuzzy. Its flowers are bilateral. It is usually found on caliche outcrops or on tight soils along drainages, and it also occurs on tight sandy loam soils in the southern portion of the Sand Sheet or on caliche ridges near hilltops and along creeks. Although it is beautiful in bloom, it can form dense thickets. It generally grows in areas where few other plants do. It is not considered a preferred browse for deer.

Figure 267. Purple sage—spring

Figure 268. Purple sage flower—spring

Texas Hogplum

Colubrina texensis (Torr. & A. Gray) A. Gray

Texas hogplum is a ubiquitous plant in the dry, sandy parts of the Sand Sheet. Its primary visual clue is the arrangement of its branches, which are delicate looking and form a distinctive lattice (see fig. 269). The blades are ovate and deeply veined, and the small, greenish-yellow flowers are star shaped. It can form dense thickets, creating prime habitat for northern bobwhites as both loafing and escape cover. It also produces black fruits containing several seeds that remain for a few months and are consumed by northern bobwhites (see fig. 271). It is not preferred deer browse.

Figure 269. Texas hogplum bush—November

Figure 270. Texas hogplum flowers—April

Figure 271. Texas hogplum fruit—January

Narrowleaf Forestiera or Elbowbush (*Panalero, Chaparral Blanco*)
Forestiera angustifolia Torr.

Narrowleaf forestiera's branches grow from its limbs at roughly ninety-degree angles, which causes the branches to intersect in an intricate fashion, and this may be the first thing an observer notices. The leaves are small and basically linear, which gives these branch patterns visibility. The berries are blue and eaten by many forms of wildlife. It is fair browse for deer.

The plant is about the same size as Texas hogplum and could be confused with it (see fig. 269). Texas hogplum also has an intricate pattern of branches, but they present themselves in a zigzag fashion.

Figure 272. Narrowleaf forestiera bush—July

Figure 273. Narrowleaf forestiera leaves and berries—June

Guayacan

Guaiacum angustifolium Engelm.

Guayacan is a small tree or large bush that is heavily draped in linear leaves. It is spectacular in the few weeks it is in bloom (usually early spring), because of the profusion of flowers covering the limbs. Its flowers have five-pointed, blue or lavender petals that tend to curl, and a highly visible set of yellow anthers. Later the showy, bright red seeds emerge from small pods. Its wood is said to be the hardest in Texas, which made it valuable for tool handles and the like. Guayacan is excellent browse for deer.

Figure 275. Guayacan leaves and flowers—March

Figure 274. Guayacan—March

Figure 276. Guayacan flower—March

Agarito or Algerita (*Palo Amarillo*)
Mahonia trifoliolata (Moric.) Fedde

Agarito is an evergreen shrub that is more common in the northern and eastern portions of the Sand Sheet. Its most prominent visual feature is its leaves, which are stiff and trifoliate and have three to five sharp points on their margins. The leaves basically look like holly leaves. In late winter, when these photos were taken, the shrub produces small yellow flowers. Later in the spring through early summer, the shrub produces small red berries that are eaten by birds. It is not eaten by deer but is excellent cover for northern bobwhites.

Figure 277. Agarito bush—February

Figure 278. Agarito flowers and leaves—February

Leatherstem (*Drago, Sangre de Drago*)
Jatropha dioica Cerv.

Leatherstem is a woody, colony-forming shrub. Its most easily recognizable feature is its stems, which are thick, reddish, and leathery or rubbery. Its leaves are spatulate. Leatherstem is avoided by grazers but produces large seeds that are favored by game birds, especially white-winged and white-tipped doves. It tends to multiply after mechanical attempts to control it. It is most common on reddish sandy loam soils.

Figure 279.
Leatherstem—October

Figure 280.
Leatherstem—
May

Peach Bush (Duraznillo)
Prunus texana D. Dietr.

Peach bush is a unique, small-statured shrub found on very sandy soils. It is endemic to Texas. Its leaves are ovate to elliptic and have prominent veins. The bark is grayish or reddish. For a short period in the winter, it produces small white flowers with five petals and long golden stamens (see fig. 283). It is often the only woody cover in very sandy open grasslands. The fruits are relished by many wildlife species.

Figure 281. Peach bush—February

Figure 283. Peach bush—February

Figure 282. Peach bush flowers—April

Texas Lantana (*Yerba de Cristo, Monte Cristo*)
Lantana urticoides Hayek

Texas lantana is one of the glories of the Sand Sheet. Its yellow, orange, and red flowers will catch the observer's eye before anything else on Sand Sheet pastures in the spring.

What appear from a distance to be large flowers shaped something like chrysanthemums are really clusters of smaller red, yellow, and orange flowers. The leaf blades are ovate to round with toothed margins. It is really impossible, at least in the Sand Sheet in spring and summer, to misidentify this plant. In fall and winter, it is a different story. When its flowers drop off, lantana becomes an unremarkable, dowdy plant that is recognizable mainly by the shape of its leaves (see fig. 286). It is occasionally eaten by both deer and cattle, and it provides good cover for northern bobwhites.

Figure 284. Texas lantana—March

Figure 285. Texas lantana—March

Figure 286. Texas lantana—November

Whitebrush (*Vara Blanca, Vara Dulce, Troncoso*)
Aloysia gratissima (Gillies & Hook.) Troncoso

Whitebrush is a tall, slender, many-branched plant that populates itself aggressively. In the spring, it has white flowers. In limited quantities, it forms valuable habitat for quail, and especially deer. If allowed to proliferate, however, it forms thickets that are virtually impenetrable to humans and cattle, and its foliage becomes so thick that nothing can grow underneath it (see fig. 288). Whitebrush is also known as bee brush—an epithet attesting to its value to bees and other pollinators.

Figure 287. Whitebrush— March

Figure 288. Whitebrush— September

Shrubby Blue Sage (*Crespa*)
Salvia ballotiflora Benth.

Shrubby blue sage is a many-branched shrub that can attain eight feet in height. Its leaves are ovate and have toothed margins. It produces small blue flowers (fig. 290). It is usually found on gravelly or rocky soils. It is not favored by deer but is excellent for pollinators, and showy when in bloom.

Figure 289. Shrubby blue sage—April

Figure 290. Shrubby blue sage leaves and flowers—April

False Mesquite
Calliandra conferta Benth.

False mesquite is a low-growing shrub that is usually less than a foot tall. It is a common plant of saline and alkaline sites and resembles a miniature mesquite or huisache. It has showy yellow blooms and infrequently produces seeds eaten by wildlife. False mesquite is an important vegetation cover on sites that grow little else.

Figure 291. False mesquite—June

Ephedra (*Popote, Canatilla, Popotillo*)
Ephedra antisyphilitica Berl. ex C. A. Mey.

Ephedra has tube-shaped limbs and produces cones resembling acorns. As the Latin name suggests, the plant was once thought to cure syphilis. It is excellent browse for deer and livestock.

Figure 292. Ephedra—January

Chile Pequin

Capsicum annuum L. var. glabriusculum (Dunal) Heiser & Pickersgill

Chile pequin is an iconic native plant of the Sand Sheet. It is a subshrub that grows from a woody perennial base, often in the canopy of mesquite or hackberry. The round green peppers that mature to red are a staple spicy ingredient in South Texas salsas, sauces, and jellies and are often served raw still on the branch. Hybrid chile pequin–like peppers or horticultural varieties of the pepper are often found growing wild as annuals near urban centers. The native chile pequin characteristically has round fruits and heart-shaped leaves. A variety of birds eat the fruits, including chachalacas and Rio Grande wild turkeys.

Because of its red berries, it can be mistaken at first glance for tasajillo in the winter (fig. 353), but a quick inspection will reveal obvious differences.

Figure 293. Chile pequin—December

Shrubs with Thorns

We have grouped together three plants that bear a confusing resemblance to each other: allthorn, lotebush, and goatbush. They are confusing because all have prominent thorns that are more like pointed limbs. Goatbush, however, is easily distinguishable from the other two by its smaller stature. It rarely grows more than about four feet high, whereas allthorn often attains six to seven feet and lotebush seven to eight.

Allthorn (Junco)
Koeberlinia spinosa Zucc.

Allthorn is so named because it is all thorns. Its thorns (spines) are long and spindly, however, and tend to bend, which distinguishes them from the thorns of lotebush and goatbush, which are stiff and straight. Allthorn's bark is smooth. Its flowers are white, its berries are reddish purple in early spring (see fig. 295) and turn black later in the year, and it is leafless (which also distinguishes it from lotebush and goatbush). Allthorn is most common in the far western and southern portions of the Sand Sheet. It is avoided by browsers.

Figure 294. Allthorn bush—April

Figure 295. Allthorn berries—April (courtesy of David Grall)

Figure 296. Allthorn blooms—March

Lotebush (*Clepe*)
Ziziphus obtusifolia (Hook. ex Torr. & A. Gray) A. Gray

Lotebush is a common shrub in the Sand Sheet, often found on tighter soils, along roads, and as part of thick chaparral. It usually grows three to six feet tall (although sometimes much taller) and has long spines, or thorns. Its thorns are gray green, slightly veined, and have leaves extending all the way to the tips. The leaves are ovate and notched along their margins, sometimes on one side only. In the winter, after the leaves are lost, the gray-green thorns begin to cause it to resemble allthorn. It is not preferred browse but is sometimes eaten by deer.

Figure 297. Lotebush—November

Figure 298. Lotebush thorns with leaves—May

Figure 299. Lotebush thorns without leaves—December

Goatbush (*Amargosa*)
Castela erecta Turp.

Goatbush, more commonly referred to in South Texas by its Spanish name, amargosa, is a low-growing shrub with the visual impression of a tangle of spines and thorns. It has small red berries that can be very showy and are somewhat similar to those of guayacan. Its leaves do not conform to a single shape and vary from linear to spatulate. A visual confirmation is the silver color on the bottom sides of the leaves. Amargosa sometimes forms near-monocultures on certain soils near drainages, often in proximity to dense stands of cenizo. The shrub can also be found in mixed brush on sandy loam sites. It is not preferred by deer.

Figure 300. Goatbush—October

Figure 301.
Goatbush—September

Figure 302. Goatbush berries
—February

LOW-GROWING THORNED SHRUBS
Wolfberry (*Cilindrillo*)
Lycium berlandieri Dunal

Wolfberry is a bush of medium height (from about three to six feet) that stands out in the winter because it is a remarkable emerald green, which contrasts with the monochromatic appearance of most other South Texas plants during that time (see fig. 305). Its leaves are linear to slightly spatulate, and its limbs are gray to reddish in some cases. It produces a red berry that is eaten by northern bobwhites and many other birds and small animals. It is found throughout the Sand Sheet, but most commonly in the western portion on sandy loam soils.

Figure 303. Wolfberry—June

Figure 305. Wolfberry —December

Figure 304. Wolfberry berries —May

Catclaw Acacia (*Uña de Gato*)
Senegalia wrightii (Benth.) Britton & Rose

Catclaw acacia is a medium-sized bush that is identifiable by its recurved thorns, visible in figure 307. Its leaves are small, rounded, and opposite. From late spring to early fall, it produces a white, brushy flower, also pictured in figure 307, and a legume. Catclaw can grow in nearly impenetrable spreads (see fig. 308), but it is an excellent cover plant for northern bobwhites, is browsed by deer and cattle, and is also important for bees and pollinators when in bloom.

Figure 306. Catclaw bushes—April

Figure 308. Catclaw colony—July

Figure 307. Catclaw flowers—April

Berlandier Acacia (*Guajillo*)
Senegalia berlandieri Britton & Rose

Guajillo is a medium-sized shrub with gray to white branches and small thorns. Its leaves are fernlike. It can form thickets on thin soils on higher-elevation sites of the Sand Sheet. The legumes (see fig. 310) produce dark brown seeds. Its leaves are browsed by white-tailed deer. It is a common component of tighter soils in South Texas. Guajillo is an important pollinator plant, and honey from bees nectaring on the plant is considered a delicacy.

Figure 309. Guajillo—March

Figure 310. Guajillo legumes—May

TALLER SHRUBS
Bluewood Condalia (*Brasil, Capul Negro*)
Condalia hookeri M. C. Johnst.

Brasil is an often irregularly shaped bush (or tree) whose limbs are obscured by a profusion of green leaves. In winter, it stands out because of its lime-green color. (The other bush that is bright green in the winter is wolfberry. Wolfberry is a small shrub with linear leaves, however, whereas brasil is a small tree.) Its leaves are broadly teardrop shaped (see fig. 313) and sometimes notched at the tips. Its limbs are gray and veined. While its berries are black at maturity, they go from yellow to red to black as they mature (see fig. 312). Northern bobwhites, turkeys, and deer eat the fruit, and deer browse the leaves. It is a very valuable woody plant for wildlife.

Its limbs are also dominated by long spines, and in some cases brasil could be confused with lotebush. But brasil leaves tend to form profuse clusters around the spines (see figs. 311 and 312), and lotebush leaves do not. Leaf shape is another way to tell brasil and lotebush apart.

Figure 311. Brasil—April

Figure 313. Brasil leaves—March

Figure 312. Brasil berries—May

Spiny Hackberry (*Granjeno*)
Celtis ehrenbergiana (Klotzsch) Liebm.

Spiny hackberry, or granjeno, as it is commonly known in South Texas, is an important and common South Texas shrub. Part of its importance to wildlife is that it produces berries (see fig. 315) later in the season than most other plants. They are eaten by many birds and mammals, including, importantly, northern bobwhites. It is a scraggly, irregularly shaped shrub with toothed, oblong/elliptic leaves. Its peculiarity is its paired thorns that grow, in cross section, at 120 and 240 degrees (in two of three equal placements, but with the third missing).

Deer commonly browse the leaves, as will cattle. In most years, following copious rainfall, mass migrations of American snout butterflies will descend on the Sand Sheet. They lay their eggs on granjeno, and several weeks later, the snout caterpillars will utterly defoliate it. The plants recover easily in time.

(At the risk of pedantry, granjeno is pronounced "gran-hay-no," not "gran-hane-yo," as many Anglo visitors to deep South Texas seem to believe.)

Figure 314. Granjeno bush—July

Figure 315. Granjeno leaves and berries—November

Lime Pricklyash (*Colima*)
Zanthoxylum fagara (L.) Sarg.

Lime pricklyash often grows in the midst of other trees (fig. 316). However, it can also grow independently into large bushes eight to ten feet tall (fig. 318). It has recurved thorns and an unusual leaf type. There are winged segments along the "rachis" (the stem between each leaf). The leaves are oblong/ovate with toothed margins, and they also sometimes curl in a convex manner along their margins (see fig. 316). The leaves give off a lime smell when crushed, and the plant is a member of the same family as citrus. The seeds are dark red to brown and grow in clusters until ripe (fig. 316). Once they fall from the plant, they are shiny and black and are consumed by northern bobwhites and white-winged doves. Lime pricklyash is also browsed by deer. Along with granjeno, brasil, and mesquite, it is one of the most common woody plants in the Sand Sheet, as well as an important woody cover plant.

Figure 316. Lime pricklyash—July

Figure 317. Lime pricklyash as a small bush—January

Figure 318. Lime pricklyash as a large bush—January

Twisted Acacia (*Huisachillo*)
Vachellia schaffneri (S. Watson) Seigler & Ebinger

Twisted acacia is a deciduous shrub with an intricate, twisted limb structure (see fig. 320). Its leaves are fernlike and it produces a dark red to black legume. Its limbs grow very close to the ground and often provide shelter for northern bobwhites. Its flowers are yellow (see fig. 321). Deer and cattle will browse the plant, and it is excellent for pollinators. Some older plants can grow into small trees. It is usually found on sandy loam soils in the Sand Sheet.

Figure 319. Twisted acacia—March

Figure 320. Twisted acacia—February

Figure 321. Twisted acacia flowers—March

Blackbrush Acacia (*Chaparro Prieto*)
Vachellia rigidula (Benth.) Seigler & Ebinger

Blackbrush acacia is a thorny shrub that can form impenetrable thickets. It has numerous limbs growing from its base. Its leaves are oblong or spatulate, and it blooms in the spring, producing profuse fuzzy, cream-colored flowers in spikes (see fig. 323). It is one of the first and most prolifically blooming woody plants. It also produces a curved, reddish legume. Northern bobwhites eat the seeds and use the plant as escape cover, as well as nesting cover. It is considered excellent browse for deer. It is typically found on tight sandy loam soils, on caliche ridges, and near drainages.

Figure 322. Blackbrush acacia—May

Figure 323. Blackbrush acacia flowers—March

Figure 324. Blackbrush acacia legumes—July

Trees

As noted in the introductory paragraph in the "Shrubs without Thorns" section, the distinction between shrubs and trees in the Sand Sheet is arbitrary in many cases. Below appear plants that probably everybody could agree are trees (although some of the plants in the shrub sections could also be called trees—for example, brasil). They are not grouped by whether they are thorned or unthorned, but rather by rough order of commonality.

Honey Mesquite (*Mesquite*)
Prosopis glandulosa Torr.

Mesquite is by far the most common tree in South Texas. It is a twisted, rough-barked, thorned tree with compound linear leaves and can grow forty feet high. It is extremely aggressive and will form thickets over large areas after overgrazing if not controlled. Even on well-managed pastures, new growth of mesquite will insinuate itself at a regular pace. Mesquite is both a bane and a blessing to managers, as it is also easily one of the most valuable plants for wildlife because of the cover and shade it provides and the bean crops that feed wildlife and cattle.

Figure 325. Honey mesquite motte—February

Figure 326. Honey mesquite flowers—May

Sweet Acacia (*Huisache, Uña de Cabra*)
Vachellia farnesiana (L.) Wight & Arn.

Sweet acacia, or huisache, as it is commonly known, is a thorned, spindly tree that favors wet soil. Its limbs grow outward from a small base, and its leaves are fernlike, typical of acacias. In the spring, it covers itself with fuzzy yellow blooms (fig. 328). It can form thickets. Huisache is the harbinger of spring in the Sand Sheet, and it is the first plant to bloom as winter wanes. Huisache is browsed heavily by deer and cattle, and the flowers are important for pollinators. Despite its value, it is an aggressive invader of grasslands, and a frequent object of control by brush managers. Huisache is more common in the eastern Sand Sheet, but it is also found in wet areas of the western portions.

Figure 327. Huisache—March

Figure 328.
Huisache
flowers—March

Sugar Hackberry
Celtis laevigata Willd.

Sugar hackberry is common along drainages in the Sand Sheet, but it can also be found on some uplands. The leaves are bright green and the bark is gray. An unmistakable key to identification is the bark, which has irregular, rough, cleat-like protrusions (fig. 330). The fruit is red to maroon and can be prolific in some years. Older mature trees are important for mast production for wildlife, as a source of shade for livestock, and as roosts for Rio Grande wild turkeys.

Figure 329. Sugar hackberry—March

Figure 330. Sugar hackberry —January

Saffron Plum (*Coma, Caimito, Coma Resimera*)
Sideroxylon celastrinum (Kunth) T. D. Penn.

Coma is a small tree appearing mainly on the western side of the Sand Sheet. Its bark is slightly grooved and gray to brown. It is usually multitrunked, and in this respect it resembles chapote (see fig. 332), but its leaves are teardrop shaped, whereas chapote leaves are oblong. It produces a dark blue to black berry in late spring. Another visual key is the pointed, thornlike tips of its twigs, but it is not considered to be a thorned tree.

It is an excellent plant for wildlife; deer browse its leaves, and all wildlife eat its fruits. When in bloom, the plant has an almost overwhelming sweet smell, and it is infested with insects and pollinators of all kinds. Coma mottes form a unique microhabitat, a feature that is desirable to many ranch owners because of the value to wildlife.

Figure 336.
Coma—May

Figure 337. Coma leaves and berries—June

Wild Olive (*Anacahuita*)
Cordia boissieri A. DC.

Wild olive is thought to be one of the most beautiful trees in South Texas, and it is often used for ornamental purposes. It is multitrunked and forms a regular, rounded shape. Its long leaves are oblong/oval, have a slight V shape in cross section, and tend to hang downward (fig. 339). Its large, five-petaled flowers are white with yellow centers. Wild olive is very common in some sandy loam sites in the southwestern Sand Sheet. It is found sporadically elsewhere.

Figure 338. Wild olive tree—May

Figure 339. Wild olive flowers—May

Texas Ebony (*Ebano*)
Ebenopsis ebano (Berl.) Barneby & Grimes

Texas ebony is taller than most South Texas trees, and its smaller limbs have an erratic, zigzag appearance. Its leaves are small, round, and opposite, giving its smaller limbs the appearance of long, uniform, double-sided rows of stacked circular shapes. This is a strong visual key when viewed up close (fig. 341). It produces long, curved seedpods. White-winged doves often use it as a nesting site, and it is an important pollinator plant. It can grow on the tops of small hills throughout the Sand Sheet, but it is more common in the western and southern portions of the area.

Figure 340. Texas ebony tree—May

Figure 341. Texas ebony leaves and legumes —May

Anaqua (*Manzanillo*)
Ehretia anacua (Terán & Berl.) I. M. Johnst.

Anaqua is also a tall tree frequently used for ornamental purposes, and its thick canopy makes it a good shade tree. Its leaves are ovate, serrated, and hairy, which makes them scratchy to the touch when rubbed backward. The plentiful round fruits are yellow to orange. The tree provides mast that is consumed by wildlife.

Figure 342. Anaqua tree—May

Figure 343. Anaqua leaves and berries—May

Jerusalem Thorn (*Retama*)
Parkinsonia aculeata L.

Retama is a common South Texas tree that can grow to twenty feet tall at maturity. Its chief visual feature is its leaf structure. Many tiny leaves grow on both sides of a long, flat, stringy rachis (see fig. 345), giving the tree a feathery, wispy overall appearance. Its flowers are yellow with red or orange centers and are usually present from late spring throughout summer. Deer browse the foliage, and northern bobwhites eat the seeds.

Figure 344. Retama tree—May

Figure 345. Retama leaves and flowers—May

Pepperbark, Hercules' Club, or Tickletongue
Zanthoxylum clava-herculis L.

Pepperbark is a small deciduous tree that looks very different with leaves and without. It has an unusual limb structure, which can be seen in the February photo (fig. 347). The limbs tend to bend upward rather than downward. It is a heavily thorned tree with large, straight thorns. Its leaves are elliptic/oval with toothed margins, and they tend to curve inward in a rounded V fashion. The tree is found only in the far eastern margin of the Sand Sheet, near the coast.

It is called "tickletongue" because its leaves cause a tickling (or numbing) sensation if chewed, and "pepperbark" because its limbs put off a strong peppery smell when cut, which some find unpleasant. It is a good wildlife cover plant, and it produces seeds eaten by game birds and other wildlife.

Figure 346. Pepperbark tree—May

Figure 347. Pepperbark tree—February

Figure 348. Pepperbark thorns and leaves—May

Cacti

Prickly Pear (*Nopal*)
Opuntia engelmannii Salm-Dyck ex Engelm.

As they say, prickly pear needs no introduction. It is pervasive over the whole of South Texas. Its pads are paddle shaped, it grows in large clumps, and it produces yellow and red flowers in the spring and summer. In late spring and summer, its flowers become fruits (fig. 351). Despite its thorns, deer, hogs, and cattle will eat it. The only cactus on the Sand Sheet that resembles it is low-growing prickly pear (see fig. 352).

Figure 349. Prickly pear—April

Figure 350. Prickly pear flowers—April

Figure 351. Prickly pear fruits—August

Low-Growing Prickly Pear
Opuntia macrorhiza Engelm.

By appearance, low-growing prickly pear is just a diminutive version of prickly pear, with one difference besides size. Its flowers are slightly different, in that they have red at their bases and only a one-petal array, whereas prickly pear's flowers have multiple levels of petals.

Figure 352. Low-growing prickly pear—April

Christmas Cactus (*Tasajillo*)
Cylindropuntia leptocaulis (DC.) F. M. Knuth

Tasajillo is also ubiquitous over the Sand Sheet. It is a bushy, many-branched cactus that produces red fruits, mostly in the summer and early fall. Its thorns are barbed and its stems are segmented, so that any abrupt disturbance of the plant causes its segments to break off and fly through the air, attaching themselves to whatever animal caused the disturbance, including humans (it is often called "jumping cactus"). In that manner, it propagates itself over long distances. Its fruits are eaten by Rio Grande wild turkeys and deer.

At first glance, chile pequin (fig. 293) can be mistaken for tasajillo because of its red fruits and its tendency to grow in the understory of larger plants. Chile pequin, however, is a subshrub, not a cactus.

Figure 353.
Tasajillo—November

Figure 354. Tasajillo
fruits—October

Horse Crippler (*Visnaga, Manca Caballo*)
Echinocactus texensis Hopffer

As with many Sand Sheet plants, the name of this cactus comes from practical observation. It is a squat, segmented cactus with formidable, rigid thorns that protrude in multiple directions. It is capable of crippling horses by penetrating the soft places in their hooves. It produces a red fruit that is said to be palatable. Its flowers are white or pink, but when not in bloom it is relatively hard to spot, which is partly why it is dangerous to animals.

Figure 355. Horse crippler—May

Strawberry Cactus (*Pitaya*)
Echinocereus enneacanthus Engelm.

Strawberry cactus is so named because it is apparently edible, with fruit that tastes like strawberries and looks like them when unopened. When in bloom, its flowers are pink. The cactus is mainly vertical and ridged, with thorns growing from the edges of the ridges. It can grow in large colonies. It is found mainly in the central and western parts of the Sand Sheet.

Figure 356. Strawberry cactus—December

Yuccas

While commonly thought to be members of the cactus family, yuccas are a group unto themselves and are thus shown here in a separate section.

Spanish Dagger (*Pita*)
Yucca treculeana Carrière

Spanish dagger is another of the signature plants of South Texas when in bloom. It grows on a trunk from which long, straight, sharply pointed leaves radiate. The leaves are thick along their spines and the edges curve inward, forming a concave cross section, a structure that contributes to their strength and stiffness. When in bloom, its flowers are cream colored to white and grow upward from the trunk in an eye-catching multiflowered column. It can grow quite tall, up to twenty or twenty-five feet, and it is often planted for ornamental purposes. The leaves are eaten by deer and cattle, and Harris's hawks use it as nesting sites.

Figure 357. Spanish dagger in bloom—February

Figure 358. Spanish dagger—September

Buckley's Yucca
Yucca constricta Buckley

Buckley's yucca is a small plant with dagger-shaped leaves that are narrow, bend easily, and grow from a base on the ground (which distinguishes it from Spanish dagger, with which it might be confused early in its growth). They have white, stringy threads on their margins (best seen in fig. 360). Buckley's yucca also produces a tall stem with white flowers when it is in bloom in the spring and summer (fig. 359), but its blooms are not as tightly bunched as those of Spanish dagger.

Figure 359. Buckley's yucca in bloom—May

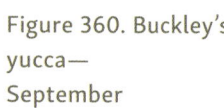

Figure 360. Buckley's yucca—September

Vines

Climbing Milkweed (*Tayalote*)
Cynanchum racemosum (Jacq.) Jacq.

Climbing milkweed is a vine that is easily recognizable in the fall and winter because of its large seedpods (about three inches) that look like peppers. They break open at maturity and release white seeds. Its leaves, which are browsed by deer, are shaped like arrowheads.

Figure 361. Climbing milkweed—November

Old Man's Beard (*Barbas de Chivato*)
Clematis drummondii Torr. & A. Gray

Old man's beard is also a climbing vine. In the spring, it has long, profuse, wispy white-to-green stamens, resembling an old man's beard. In the summer, it seems to be everywhere, mostly draped on fences.

Figure 362. Old man's beard—June

Figure 363. Old man's beard—August

Variable Leaf Snailseed (*Correhuela*)
Cocculus diversifolius DC.

Correhuela is a vine that twines around the limbs of host plants (a catclaw in fig. 364). It produces small, five-petaled flowers and purple fruits.

Figure 364. Correhuela—September

Tievine

Ipomoea cordatotriloba Dennst.

Tievine is a twining vine having pink flowers with purple centers that bloom in the summer and fall. Its flowers somewhat resemble poppies. Its leaves are deeply lobed.

Figure 365. Tievine—August

Mistletoe (*Injerto*)
Phoradendron tomentosum (DC.) Engelm. ex A. Gray

While not technically a vine, mistletoe is a common parasitic plant found in trees in the Sand Sheet. Its leaves are elliptic, and it has small white flowers when in bloom. Its most recognizable feature is its green mass, engulfing tree limbs that might otherwise be leafless. It prefers mesquite trees and can eventually kill its host.

Figure 366--. Mistletoe—November

Glossary

Alternate—with leaves emanating from the stem alternately, right-left, right-left.

Annual—a plant that completes its life in a single year or growing season.

Anther—the head of a filament (the stalk that, together with the anther, forms the stamen of a flower); the pollen-bearing part of the flower.

Awns—hairs that emanate from seeds.

Bilateral—longitudinally symmetrical; both longitudinal halves are identical.

Bracts—small, pointed structures, technically leaves, that are interspersed with flowers in an inflorescence. Bracts essentially support the flower.

Bunch—a group or clump of plants in which each individual plant is distinct.

Calyx—the underside base of a flower, made up of sepals.

Colony—a large mass of plants in which individual plants are indistinguishable.

Compound—with two or more leaves growing from the same node.

Corolla—the petals of a flower as a whole.

Disk flowers—the components of the head of a sunflower, technically themselves flowers.

Elliptic—pointed at both ends and wide in the middle, about one-third as wide as long.

Endemic—a plant that is found nowhere else in the world.

Filament—the stalk of a stamen.

Head—a collective term for the flowering structure of a plant, such as the flowers, or the seeds of a grass.

Inflorescence—a flowering cluster, or collectively, the entire reproductive structure that bears seeds or fruits.

Lanceolate—resembling the tip of a lance; rounded at the base, sharp and pointed at the tip, and narrow.

Linear—long and narrow, with parallel sides.

Lobed—with significant indentations along the edges, like oak leaves.

Oblanceolate—narrow; similar to lanceolate, but pointed at both ends.

Oblong—with parallel sides and rounded, like the link of a chain; similar to lanceolate, but wider.

Obovate—egg shaped, with the broader part at the end; the opposite of ovate.

Opposite—with leaves emanating from the stem opposite each other, that is, growing from the same point on the stem across from each other.

Ovate—egg shaped, with the narrower part at the end; the opposite of obovate.

Palmate—separating like fingers from the palm of a hand.

Perennial—a plant that lives more than two years.

Pistil—the seed-producing female part of a flower, found at the center of the corolla and surrounded by stamens.

Rachis—the main stem between each leaf stem.

Ray flowers—the petals of a sunflower or typical flower.

Sepals—the bladelike extensions from the stem underneath the corolla, making up the calyx.

Spatulate—narrow at the base and wider and round at the tip, but not as wide as obovate; "spatulate" is basically the opposite of "lanceolate."

Stamen—the male part of a flower; a rod-shaped protrusion emanating from the head of the corolla. Stamens consist of filaments and anthers.

Toothed—with toothlike serrations along the edges.

Bibliography

Carr, W. R. 2018. "Some Plants of the South Texas Sand Sheet." University of Texas at Austin. Accessed July 2, 2018. http://w3.biosci.utexas.edu/prc/DigFlora/WRC/Carr-SandSheet.html.

Correll, D. S., and M. C. Johnston. 1970. *Manual of the Vascular Plants of Texas*. Renner: Texas Research Foundation. 1881 pp.

Everitt, J. H., D. L. Drawe, C. R. Little, and R. I. Lonard. 2011. *Grasses of South Texas: A Guide to Identification and Value*. Lubbock: Texas Tech University Press. 321 pp.

Everitt, J. H., D. L. Drawe, and R. I. Lonard. 1999. *Field Guide to the Broad-Leaved Herbaceous Plants of South Texas Used by Livestock and Wildlife*. Lubbock: Texas Tech University Press. 277 pp.

———. 2002. *Trees, Shrubs, & Cacti of South Texas*. Rev. ed. Lubbock: Texas Tech University Press. 249 pp.

Gould, F. W. 1975. *The Grasses of Texas*. College Station: Texas A&M University Press. 653 pp.

Gould, F. W., and T. W. Box. 1965. *Grasses of the Texas Coastal Bend*. College Station: Texas A&M University Press. 189 pp.

Hatch, S. L., J. L. Schuster, and D. L. Drawe. 1999. *Grasses of the Texas Gulf Prairies and Marshes*. College Station: Texas A&M University Press. 355 pp.

Jones, F. B. 1982. *Flora of the Texas Coastal Bend*. Sinton, TX: Welder Wildlife Foundation. 267 pp.

Lady Bird Johnson Wildflower Center. 2018. Native Plants Database. Accessed July 2, 2018. http://www.wildflower.org.

Lehman, R. L., R. O'Brien, and T. White. 2005. *Plants of the Texas Coastal Bend*. College Station: Texas A&M University Press. 352 pp.

Richardson, A. 2002. *Wildflowers and Other Plants of the Texas Beaches and Islands*. Austin: University of Texas Press. 247 pp.

Richardson, A., and K. King. 2011. *Plants of Deep South Texas*. College Station: Texas A&M University Press. 457 pp.

Shaw, R. B. 2012. *Guide to Texas Grasses*. College Station: Texas A&M University Press. 1080 pp.

Taylor, R. B., J. Rutledge, and J. G. Herrera. 1997. *A Field Guide to Common South Texas Shrubs*. Austin: Texas Parks and Wildlife Press and University of Texas Press. 106 pp.

Weniger, D. 1984. *Cacti of Texas and Neighboring States*. Austin: University of Texas Press. 356 pp.

University of Texas at Austin. 2018. The Lundell Plant Diversity Portal. Billie L. Turner Plant Resources Center. Accessed July 2, 2018. https://prc-symbiota.tacc.utexas.edu.

USDA Natural Resources Conservation Service. 2018. PLANTS Database. Accessed July 2, 2018. http://plants.usda.gov.

Index

Webb Germplasm, 159. *See also* seed
source
wedgeleaf prairie clover, 110
Welder Germplasm, 162. *See also* seed
source
western ragweed, 44, 76, 115
whiplash pappusgrass, 159, 169
whitebrush (troncoso, vara blanca,
vara dulce), 181
white flowers, 40–55, 63, 77, 106, 109,
111, 116, 118, 119, 179, 181, 186, 190,
203, 211, 213, 214, 219
whitemouth dayflower, 57, 71
white-tailed deer, 7, 9, 10, 12, 18, 19, 20,
27, 37, 41, 45, 49, 56, 68, 72, 80, 92, 93,
103, 110, 113, 116, 117, 118, 175, 176, 180,
181, 184, 187, 188, 191, 192, 193, 195, 196,
198, 200, 202, 206, 208, 210, 213, 215
white-winged doves, 114, 178, 194, 204
wild buckwheat, 55
wild olive (anacahuita), 203
wild petunia, 68
wildlife, 5, 18, 21, 31, 38, 44, 50, 55,
58, 85, 86, 87, 89, 94, 100, 101, 102,
104, 105, 106, 107, 112, 122, 123, 127,
128, 132, 133, 134, 136, 149, 150, 153,
159, 161, 162, 163, 164, 167, 169, 172,
175, 179, 183, 192, 193, 197, 199, 200,
201, 202, 205, 207. *See also* deer;
hogs; northern bobwhite; quail; Rio
Grande wild turkeys
Wilman lovegrass, 153
windmillgrasses, 161
winecup, 67
wolfberry (cilindrillo), 189, 192
woodland sensitive pea, 5
woolly croton, 94, 96
woolly globemallow, 91, 92
woolly tidestroma, 116
woolly white, 54, 55

yellow berries, 192, 205
yellow flowers, 5–39, 70, 85–87, 100,
101, 102, 107, 108, 122, 123, 127, 174,
177, 180, 183, 195, 198, 201, 206, 208,
209
yellow Indiangrass, 128, 134
yerba de cristo (monte cristo, texas
lantana), 180
yuccas, 213

Zapata Germplasm, 82. *See also* seed
source
zizotes milkweed, 52

Other Books in the Perspectives on South Texas Series

African Americans in South Texas History
Bruce A. Glasrud

Beef, Brush, and Bobwhites: Quail Management in Cattle Country
Fidel Hernández and Fred S. Guthery

Ecología y Manejo de Venado Cola Blanca
Timothy E. Fulbright and J. Alfonso Ortega-Santos

Nesting Birds of a Tropical Frontier: The Lower Rio Grande Valley of Texas
Timothy Brush

Petra's Legacy: The South Texas Ranching Empire of Petra Vela and Mifflin Kenedy
Jane Clements Monday and Francis Brannen Vick

Plants of Deep South Texas: A Field Guide to the Woody and Flowering Species
Alfred Richardson

Racial Borders: Black Soldiers along the Rio Grande
James N. Leiker

Texas Quails: Ecology and Management
Leonard A. Brennan and Katharine Armstrong

White-Tailed Deer Habitat: Ecology and Management on Rangelands
Timothy E. Fulbright and J. Alfonso Ortega-Santos

Upland and Webless Migratory Game Birds of Texas
Leonard A. Brennan, Damon L. Williford, Bart M. Ballard, William P. Kuvlesky, Eric D. Grahmann, and Stephen J. DeMaso

Wildlife Ecology and Management in Mexico
Edited by Raul Valdez and J. Alfonso (Poncho) Ortega-S.